# ASSESSMENT OF THE INCENTIVES
# CREATED BY PUBLIC DISCLOSURE
# OF OFF-SITE CONSEQUENCE ANALYSIS INFORMATION
# FOR REDUCTION IN THE RISK OF ACCIDENTAL RELEASES

April 18, 2000

United States Environmental Protection Agency
Ariel Rios Building
1200 Pennsylvania Avenue, NW
Washington, DC 20460

This page intentionally left blank.

# TABLE OF CONTENTS

# LIST OF TABLES

# LIST OF EXHIBITS

# EXECUTIVE SUMMARY

Under the Chemical Safety Information, Site Security and Fuels Regulatory Relief Act, the President delegated to the Administrator of the U.S. Environmental Protection Agency (EPA) the task of assessing the incentives for reduction in accidental chemical releases created by public disclosure of off-site consequence analysis information. This document reports the results of EPA's assessment.

In the wake of the chemical tragedy in Bhopal, India, and a series of large chemical accidents in the U.S. in the late 1980s, Congress added new provisions to the Clean Air Act for the prevention of accidental chemical releases. In particular, Congress directed EPA to require facilities that pose the greatest risk of harm to the public and the environment as a result of an accidental chemical release prepare and submit risk management plans (RMPs). An RMP must describe the facility's chemical accident prevention program, emergency response program, and off-site consequence analysis (OCA). The OCA must evaluate the potential for hypothetical worst-case and alternative accidental release scenarios to harm the public and environment around the facility. Congress mandated that RMPs be available to state and local governments and the public.

EPA promulgated RMP requirements in June 1996; the first RMPs were due three years later. To reduce paperwork burden and to take advantage of today's technology, EPA designed software tools and forms so that all RMPs could be submitted electronically to EPA and stored in a central information system. All levels of government would have immediate access to the system and the most recent RMP submissions. The vast majority of RMPs have been submitted electronically to EPA.

EPA originally planned to place the RMP information system on the Internet for easy access by the public, as well as by governments, based on the recommendation of many members of a Subcommittee created under the Federal Advisory Committee Act. However, concerns were raised that Internet access to a large, searchable database of OCA results could be used as a targeting tool by terrorists and other criminals. Although EPA subsequently decided not to place the OCA sections of RMPs on the Internet, new concerns were raised that recent amendments to the Freedom of Information Act (FOIA) would compel EPA to release this information in electronic format. Congress responded by passing the Chemical Safety Information, Site Security and Fuels Regulatory Relief Act (CSISSFRRA), which the President signed on August 5, 1999.

CSISSFRRA temporarily exempts OCA information from public disclosure under FOIA. It requires the President to "assess the increased risk of terrorist and other criminal activity associated with the posting of [OCA] information on the Internet, and the incentives created by public disclosure of OCA information to reduce the risk of accidental chemical releases." Based on these assessments, the President is to issue regulations "governing the distribution of [OCA] information in a manner that, in the opinion of the President, minimizes the likelihood of accidental releases and [any increased risk of terrorist activity associated with Internet posting of

OCA information] and the likelihood of harm to public health and welfare."

## FINDINGS OF INCENTIVES ASSESSMENT

- **Public disclosure of RMPs including OCA information would likely lead to significant reduction in the number and severity of accidental chemical releases. In addition, widespread access to OCA information by all stakeholders would serve the function Congress originally intended in the Clean Air Act Amendments — to inform members of the public and allow them to participate in decisions that affect their lives and communities. The public is not likely to generate such information on its own, and thus the greater the public access to OCA information, the more likely potential public safety benefits would be realized.**

- **Multiple segments of the public, particularly citizens, citizens' groups, and the media**, are likely to become more interested in chemical safety and chemical release risk reduction, to the extent they become aware of the potentially large consequences associated with worst-case scenarios and, to a lesser extent, alternative release scenarios. The interest and concern about potential consequences will likely trigger comparisons and detailed analyses of not only OCA information but safety and environmental performance of facilities as well. Widespread awareness of the comparisons and analyses would likely lead industry to make changes and would stimulate dialogue among facilities, the public, and local officials to reduce chemical accident risks.

- **Chemical accidents continue to impose considerable costs** in terms of human lives and health, property damage, and public welfare. Facilities covered by the RMP rule reported that from mid-1994 to mid-1999 there were about 1,900 serious accidents that caused 33 deaths, 8,300 injuries, and the evacuation or sheltering of 221,000 people. These accidents cost the affected facilities more than $1 billion in direct damages and two to four times that in business interruption losses. Almost 80% of these accidents occurred at facilities already subject to the OSHA process safety management standard, which is designed to reduce accidents. These accidents also represent less than 10 percent of all unintended releases of hazardous substances reported to the government during this period. Additional efforts are needed in order to reduce the number and severity of chemical accidents.

- **Given the opportunity, the public uses hazard information to take action that leads to risk reduction.** Various segments of the public have strong incentives to use OCA information in ways that reduce risk. For example, there is a broad consensus that national publication of the Toxics Release Inventory (TRI) data by the government, followed by analysis by citizens' groups and the news media, led to action by industry to reduce emissions. Nationally, reported TRI emissions have fallen 43 percent since 1988, a time in which industrial production has risen 28 percent. Although other factors likely contributed to the decline in emissions, negative press coverage directed at certain facilities appears to have led these facilities to achieve reductions in their TRI emissions. It is not possible to quantify the exact level of risk reduction that would be gained from public dissemination of OCA information, but the effect would likely be significant.

- **Ease of access to information is important to public use and risk reduction.** Data available in paper form on request from state or local agencies are rarely sought. For example, data on the location and identity of hazardous chemicals are requested about 3,500 times a year from Local Emergency Planning Committees (LEPCs). (There are about 3,200 LEPCs in the country and about 560,000 facilities subject to requirements to report information on hazardous chemicals to LEPCs.) Meanwhile, environmental data on Environmental Defense's "Scorecard" website are at least 250 times more likely to be reviewed by the public than information from LEPCs. Likewise, early indications are that the meetings which facilities were required to conduct by CSISSFRRA to explain OCA information to the public have drawn very few attendees, even when citizens received individual invitations. In contrast, when industry has gone out to places the public already frequents (for example, a shopping mall) and provided consequence information directly to citizens, outreach and communication about chemical accident risks have been more successful.

- **Information that puts hazards into context, as OCA data do, is far more likely to be used by the public than "raw" data.** The importance of such "interpreted" information (already analyzed in order to be understandable) is demonstrated by the increased use of TRI data when they were made available as part of Scorecard on the Internet. Although TRI data are available electronically through EPA's Envirofacts and the RTK-Net (Right-To-Know Network) websites, Scorecard ranks each facility on various indicators by county, state, and nation, and explains the health effects of chemicals emitted by that facility. The raw TRI data on RTK-Net were drawing 240,000 searches a year; Scorecard draws over a half million page views per month.

- **Although OCA data could be derived from other available data, the public is unlikely to do so.** Derivation of OCA data requires some technical knowledge and time. While motivated and skilled individuals and organizations can use widely available existing data, guidance, and models to estimate off-site consequences with relative ease, evidence suggests that the general public is unlikely to be able and willing to do so.

- **A complete RMP containing OCA information is necessary to understand the extent of the hazard** posed by a particular facility in comparison to other facilities in an area, within an industrial sector, or handling the same chemicals. While the OCA data address the hazard, the RMP information addresses the steps to control those hazards. Understanding the extent of a hazard and how it is controlled leads to understanding the risk posed by that facility.

- **The penalties for disclosure contained in CSISSFRRA are having a chilling effect,** even though the statute provides for access to OCA information for state and local officials, including emergency planners and responders, and allows those officials to communicate OCA data to the public. Many of these officials are not willing to obtain or to communicate the data and thereby to risk accidental or inadvertent disclosure of OCA information, even though CSISSFRRA penalizes only its willful disclosure. More fundamentally, making the provision of OCA data to the public discretionary leaves in the hands of government the decision about whether and to what extent to convey the data. CSISSFRRA also allows facilities to release their OCA information to the public, but that, too, is at their discretion. CSISSFRRA's requirement for facilities to conduct a public meeting or post a public notice summarizing OCA information provided only a one-time opportunity to learn about local hazards.

- **Actual chemical releases are different from the releases evaluated for OCA purposes.** No one can control all of the conditions (for example, weather) used to develop an off-site consequence analysis; actual conditions at a facility can vary widely from those used in the analysis. The accident prevention rule requires facilities to conduct OCAs in a specified, systematic manner so that the public and others can understand the relative hazards and risks posed by facilities as a result of the type and amount of chemical handled and the mitigation measures used.

This assessment finds that convenient public access to OCA information has the power to reduce real impact associated with chemical accidents. America needs to be further educated about chemical risks. Dissemination of OCA information could make an important contribution to a public dialogue about risk reduction and protection of lives. This public dialogue among community members, emergency planners and responders, and facilities at the local level is key to risk reduction.

# CHAPTER 1

# INTRODUCTION

*"I know of no safe depository of the ultimate powers of society but the people themselves; and if we think them not enlightened enough to exercise their control with a wholesome discretion, the remedy is not to take it from them, but to inform their discretion"* (Thomas Jefferson, letter to William Charles Jarvis, September 28, 1820).

The federal government's efforts to prevent and mitigate chemical accidents have come largely in the wake of the 1984 accidental chemical release in Bhopal, India, that killed more than 3,000 people and injured more than 100,000 (1). This incident demonstrated to the world the magnitude of the potential consequences of a single chemical accident.

But this was not an isolated event. Less than one month prior to the Bhopal accident, an accidental release of liquefied petroleum gas (i.e., propane) from a storage terminal in Mexico City resulted in a large fire and series of explosions, killing 500 people and destroying a residential area (2). Other catastrophic chemical accidents have occurred in countries throughout the world, including the United States. In 1985, an accident at a Union Carbide plant in Institute, West Virginia, led to a release of a noxious mixture of methylene chloride and aldicarb oxime, resulting in the hospitalization of 134 people living in surrounding areas.

As a result, Congress passed the Emergency Planning and Community Right-to-Know Act (EPCRA) in 1986 as a part of the Superfund Amendments and Reauthorization Act. EPCRA calls on states to create State Emergency Response Commissions (SERCs) and communities to form Local Emergency Planning Committees (LEPCs) to prepare local emergency response plans for chemical accidents. EPCRA also requires facilities to provide LEPCs with information necessary for emergency planning, and to submit to SERCs, LEPCs and local fire departments annual inventory reports and information about hazardous chemicals. The statute also established the Toxics Release Inventory (TRI), which requires certain facilities to annually report the quantities of their emissions of toxic chemicals. These data are to be available to the public and EPA is to maintain a national database containing these toxic chemical release reports.

However, EPCRA contains no provisions for the prevention of chemical accidents and, because major accidental releases continued to occur, Congress included two provisions in the Clean Air Act (CAA) Amendments of 1990 to institute federal regulatory programs to prevent chemical accidents that harm workers, the public and the environment. Section 304 of the Amendments calls for chemical accident prevention and emergency response regulations to protect workers on-site, while section 112(r) of the amended CAA calls for regulations to prevent and respond to chemical accidents that could affect the public and environment off-site (3).

5

In Section 112(r), Congress established a general duty on facilities handling extremely hazardous chemicals to do so safely (section 112(r)(1)), and required EPA to establish regulations to ensure that facilities that pose the greatest risk develop and implement chemical accident prevention and detection programs (section 112(r)(7)). Congress further directed that the chemical accident prevention regulations require that facilities prepare and submit risk management plans (RMPs); these plans must include a hazard assessment that estimates the potential consequences of hypothetical worst-case releases, an accident history, a program for preventing accidental releases, and an emergency response program (section 112(r)(7)(B)(ii)). Finally, Congress required that these RMPs be submitted to the federal Chemical Safety and Hazard Investigation Board, state and local emergency response officials, and be made available to the public (section 112(r)(7)(B)(iii)).

EPA issued a rule in 1994 that lists the most potentially acutely hazardous toxic and flammable substances along with a threshold quantity for each. In 1996 the Agency issued a rule requiring every facility handling more than the threshold quantity of a listed substance to develop and implement a risk management program based on an assessment of the hazards at that facility (the "RMP rule"). As required by section 112(r), EPA specified in the rule that the hazard assessment include an off-site consequence analysis (OCA) of the potential consequences of worst-case and alternative scenario chemical releases and that the results of the OCA be reported in the facility's RMP.

The OCA provides a rough estimate of the potential consequences to a surrounding community of one or more hypothetical accidental releases, without evaluating the likelihood or probability of such an accident occurring. Potential consequences are expressed in terms of potentially exposed population, as well as the types of buildings, parks, and other public and environmental areas that could be seriously affected by a release.

**Chemical Emergency Preparedness and Prevention - Legislative and Regulatory History**

1986 - Emergency Preparedness and Community Right-to-Know Act (EPCRA); PL99-499

1987 - Extremely Hazardous Substances List and emergency planning and reporting requirements (40 CFR 355 and 370).

1988 - Toxic Release Inventory (TRI) reporting requirements (40 CFR 372).

1990 - Clean Air Act (CAA) Amendments, containing Sections 112(r) and 304; PL101-549

1992 - Occupational Safety and Health Administration (OSHA) Process Safety Management Standard (PSM); (29 CFR 1910.119)

1994 - EPA List of Substances and Threshold Quantities for accident prevention program (40 CFR 68.130).

1996 - EPA Accidental Release Prevention Requirements: Risk Management Program (40 CFR 68).

Rather than impose new requirements for specific accident prevention measures, EPA chose to rely in part on the public availability of RMPs, including the OCA information in RMPs, to help ensure that facilities take all reasonable steps to reduce their risk of accidental releases. For many facilities covered by pre-existing accident prevention and response rules or voluntary industry standards, the requirements to conduct an OCA and prepare a publicly available RMP containing certain data elements from the OCA may be the only significant additional regulatory requirements under the RMP rule.[1]  A complete description of the RMP elements, how an OCA is conducted, and the various elements of OCA information contained in an RMP is available in **Appendix A**.

The Agency decided that all RMPs would be submitted to EPA, which would handle dissemination to state and local officials and the public.  The Agency believed that this approach would enhance dissemination and use of the RMP information.  EPA's past experience in implementing EPCRA had shown that many state and local officials needed assistance in managing the chemical information submitted to them on paper by industry under that law, and that the public often did not take advantage of this information since it was not conveniently available.

With the help of the Accident Prevention Subcommittee[2], EPA designed an RMP reporting form that lent itself to the creation of an electronic database.  The form consists of an Executive Summary and sections for reporting OCA results, prevention program data, and other information.  In the Executive Summary, reporting facilities are required to explain in prose the facility's risk management program, including a brief summary of the facility's OCA.  The remaining RMP sections, including the OCA sections, are in check-off box, yes/no and other formats that allow compilation of an electronic database.  As a result, the information in those sections is relatively general in nature (e.g., the form calls for the facility to identify the types of prevention devices it uses in a chemical process, but not where they are used or how many are used).  The vast majority of RMPs submitted by June 21, 1999, were submitted electronically to EPA.  The Agency developed and maintains a central database of RMPs from which immediate access can be provided to stakeholders who are designated recipients of the information as mandated by Congress.

To satisfy the section 112(r) requirement that RMPs be made available to the public, nearly all members of the Accident Prevention Subcommittee recommended that EPA place

---

[1] The list and RMP rules are codified at 40 CFR Part 68.

[2] The Accident Prevention Subcommittee to the Clean Air Act Advisory Committee was established to provide the Chemical Emergency Preparedness and Prevention Office (CEPPO) with stakeholder advice and counsel on scientific and technical aspects of its programs.  The Subcommittee considers technical issues, methodologies, and/or products which CEPPO provides for review.  These form the basis for Subcommittee findings and recommendations which enable CEPPO to strengthen its technical program and specific technical products.  The Subcommittee is made up of representatives from industry, state and local government, public interest groups, academia, trade associations, and professional organizations.  The Clean Air Act Advisory Committee and the Accident Prevention Subcommittee were created under the Federal Advisory Committee Act.

RMPs on the Internet for easy access by the public. However, concerns were raised that Internet access to a large, searchable database of OCA results would provide a targeting tool for terrorists and other criminals. In response, EPA decided not to place the OCA sections of RMPs on the Internet, but concerns were next raised that recent amendments to the Freedom of Information Act (FOIA) would compel EPA to release these sections in electronic format. Congress responded by passing Chemical Safety Information, Site Security and Fuels Regulatory Relief Act (CSISSFRRA), which the President signed on August 5, 1999.

CSISSFRRA temporarily exempts OCA information from public disclosure under the CAA and FOIA. It requires the President to assess the increased risk of terrorist and other criminal activity associated with the posting of OCA information on the Internet, and the incentives created by public disclosure of OCA information to reduce the risk of accidental chemical releases. Based on the assessments, the President is to issue regulations governing the distribution of OCA information in a manner that, in the opinion of the President, minimizes the likelihood of accidental releases and any increased risk of terrorist activity associated with Internet posting of OCA information and the likelihood of harm to public health and welfare.

The President delegated to the Department of Justice (DOJ) and the EPA authority to perform the required assessments and to promulgate the required regulations. The President has delegated authority to perform the assessment of the increased risk to DOJ, and has delegated authority to perform the assessment of the incentives to reduce risk to EPA. The President also jointly delegated to DOJ and EPA his duty to promulgate the regulations, subject to review and approval by the Office of Management and Budget. (For a detailed description of CSISSFRRA and the kinds of data and information available to the public and other stakeholders under CSISSFRRA or by other means, see **Appendix B**.) This document reports the results of EPA's assessment.

## ORGANIZATION OF THIS REPORT

Because RMPs were submitted only in mid-1999 and because the vast majority of OCA information is not currently available, it is not possible to analyze the impact of either complete RMPs or OCA information directly at this time. EPA, therefore, has examined other programs that provide the public with similar information related to risk. The assessment uses data from these other programs as well as from the RMP program to answer a series of questions:

*Chapter 2: Are Chemical Accidents a Serious Problem?*

Before considering the risk reduction potential of public information, Chapter 2 addresses the fundamental question of whether chemical accidents present a serious risk to the public, employees of facilities, and the environment, and whether current regulatory programs are in place to reduce the risk sufficiently. If chemical accidents pose little threat because existing programs have already reduced risks, then there is less need or benefit to be gained by making information available to the public for further risk reduction.

*Chapter 3: Does Public Information Lead to Risk Reduction?*

Chapter 3 examines the evidence that public information leads to risk reduction. This question has two parts. Does the public use the data; and does that use lead to risk reduction? In some cases, merely publicizing the data stimulates industry to take action. Chapter 3 examines data from the Toxics Release Inventory (TRI) program, accidental releases, and non-environmental programs.

*Chapter 4: Does the Type of Information and Access Make a Difference?*

Chapter 4 examines whether interpreted data, such as OCA data, are more likely to be used by the public than raw data. Interpreted data are data that are easily understood by the user without the need for further manipulation or supplemental information. Chapter 4 then examines whether ease of access to the data increases the likelihood that data will be used by the public.

*Chapter 5: Are There Other Sources of the Same Data?*

This chapter examines whether OCA data can be obtained from other sources and whether the availability is sufficient to lead to risk reduction.

*Chapter 6: How Much Information is Necessary to Spark Risk Reduction Efforts?*

This chapter examines ways OCA information could be made available and ways facilities could be categorized or grouped by the hazards or risks of accidental release they present in order to disseminate OCA information.

*Chapter 7: What is the Public's Access to OCA Information under CSISSFRRA?*

Chapter 7 describes OCA information in more detail, and examines the options for data access under CSISSFRRA.

*Chapter 8: Findings*

Chapter 8 summarizes the findings and presents EPA's conclusions with respect to this assessment.


In addition to the main body of the assessment, the report includes a number of appendices for readers interested in additional details:

Appendix A    Provides a detailed description of the RMP data, including the OCA data and discusses what is included in OCA data and what is not. A sample RMP is included.

Appendix B    Provides a detailed description of CSISSFRRA and its provisions that relate to this assessment.

Appendix C    Presents summaries of actions taken as the result of public environmental data.

Appendix D    Presents details of a study of the effects of negative press on TRI emissions.

Appendix E    Presents details of the accident data discussed in this report.

Appendix F    Discusses the individuals and groups that are likely to use OCA information, and how various uses can create or affect incentives for risk reduction.

# CHAPTER 2

# ARE CHEMICAL ACCIDENTS A SIGNIFICANT PROBLEM?

Congress enacted section 112(r) of the CAA to reduce the number and severity of accidental releases of chemicals that could cause serious harm. Although most localities are well prepared for chemical emergencies, sudden accidental releases to the air that rapidly migrate off-site (or, in the case of a flammable material, quickly reach an ignition source) potentially expose the public or environment to harmful effects in a short time. As evidence of this, sheltering-in-place has become a preferred emergency response strategy because air releases move too fast to make evacuation a feasible option. Consequently, protection of public health depends on preventing the releases.

Chemical accidents continue to be a serious problem in the U.S., causing deaths, injuries, serious property damage, and disrupting business and the lives of individuals in the vicinity of facilities. The facilities subject to the RMP rule submitted information on all of their serious accidents that occurred in the five years prior to the date of submission of the RMP (approximately June 1999).[3] Serious accidents are defined in the rule as those that cause deaths or injuries on- or off-site; significant property damage on-site; or known property damage, evacuations or sheltering-in-place, or environmental damage off-site. Overall, 1,086 facilities reported 1,913 accidents in their RMPs.

The impacts of the releases reported in the RMPs are shown in **Table 1**. It should be noted that not all of the 1,913 accidents had one or more of the impacts that require reporting; a few facilities chose to report all their releases of regulated substances rather than limit the reports to those that were subject to reporting under the rule. No off-site deaths were reported in the RMPs.

---

[3] These data cover the five-year period prior to the submission of the RMP, but because each RMP has a unique submission date, they may not cover exactly the same five years. However, most submitters probably included accidents that occurred between mid-1994 and mid-1999. Therefore, the 1994 plus the 1999 numbers are equivalent to one year's releases.

**Table 1 – Five-Year Accident History Data from RMP Submissions**

| Year | Deaths On-site | Injuries On-site | Hospitalized | Other Medical Treatment | Evacuated | Sheltered in Place | Damage ($ millions) |
|---|---|---|---|---|---|---|---|
| 1994 (partial) | 6 | 239 | 46 | 135 | 3623 | 4,396 | 356 |
| 1995 | 2 | 433 | 103 | 4,823 | 8,677 | 21,978 | 67 |
| 1996 | 4 | 369 | 28 | 334 | 2,616 | 41,799 | 129 |
| 1997 | 5 | 416 | 11 | 583 | 7,267 | 65,041 | 218 |
| 1998 | 3 | 394 | 17 | 136 | 5,723 | 52,717 | 94 |
| 1999 (partial) | 13 | 124 | 12 | 27 | 1,937 | 5,549 | 153 |
| Total: | 33 | 1,975 | 217 | 6,038 | 29,843 | 191,480 | 1,018 |

However, two recent accidents, one at a small chemical plant (4) and one at an ice-making plant, have caused or contributed to off-site deaths. The ammonia release at the ice plant is the first release of a toxic chemical in the U.S. known to have contributed to a fatality off-site as a result of exposure to the chemical, rather than impact from an explosion (**see box**).

**Table 2** assigns a dollar value to these impacts, based on the values EPA used in the Economic Impact Analysis (EIA) for the final RMP rule in May 1996. Lost production is valued conservatively, as it was in the EIA, at two times the value of property damage; according to Marsh and McLennan[4], the standard insurance industry assumption is that business losses are four times the cost of property damage (5).

---

[4] Marsh and McLennan represents the J&H Marsh and McLennan Corporation, parent company to M&M Protection Consultants (M&MPC). M&MPC underwrites risk and provides consultation to management on hazard control; they produce reviews of large property losses in the chemical and petrochemical industries.

**Table 2 – Dollar Values of Impacts of Releases in the RMP Five-Year Accident History**

| Impact | RMP Data | Unit Value[5] | Total |
|---|---|---|---|
| Deaths | 33 | $5,400,000 | $178,200,000 |
| Hospitalizations | 217 | $19,000 | $4,123,000 |
| Other medical treatment | 6,038 | $200 | $1,207,600 |
| Evacuation | 29,843 | $290 | $8,654,470 |
| Sheltered | 191,480 | $30 | $5,744,400 |
| Property Damage | $1,018,000,000 | -- | $1,018,000,000 |
| Lost Production | $2,036,000,000 | -- | $2,036,000,000 |
| Total | | | $3,252,000,000 |
| 5-Year Annual Average | | | $650,000,000 |

As costly as the accidents reported in the RMPs have been, they do not represent most of the chemical accidents in the country. RMP facilities reported 1,913 accidents for a five-year period. Over that time period, more than 25,000 hazardous substance releases were reported to EPA and the National Response Center. Because releases reported to the National Response Center are generally called in while the release is occurring when impact data are often incomplete, it is not possible to estimate the impacts or the cost of these releases. The U.S. Chemical Safety and Hazard Investigation Board maintains a Chemical Incident Reporting Center database of information about chemical releases gathered from news accounts and other incident reports (on the Internet at www.csb.gov/circ/). An informal review of incidents contained in this database shows that from May 1999 to April 17, 2000, about 350 chemical incidents occurred at fixed industrfacilities and not related to transportation. These incidents generated: 39 fatalities (employees, first responders, and one citizen off-site); more than 1,000 worker and first responder injuries; evacuations of more than 19,000 people (employees and residents); and more than $7.4 million in damages. It is clear that, whatever progress has been made to prevent accidental releases, chemical accidents continue to be a serious problem.

## CAN ACCIDENTS BE PREVENTED BY REGULATIONS ALONE?

In 1990, Congress mandated that both EPA and the Occupational Safety and Health Administration (OSHA) issue rules to prevent chemical accidents. As noted in Chapter 1, OSHA promulgated the process safety management (PSM) standard in February 1992; facilities were to be in compliance with all elements except the process hazard analysis (PHA) by 1994.[6] The PSM

---

[5] Unit values are taken from EPA's Economic Analysis in Support of Final Rule on Risk Management Program Regulations for Chemical Accident Release Prevention, as Required by Section 112(r) of the Clean Air Act, May 21, 1996.

[6] The process hazard analysis requirement was phased in over five years for facilities with multiple covered processes.

14

standard is based on the concept that managing chemical accident risk requires an integrated approach that involves identifying and assessing risks, managing risk through the adoption of practices (such as operating procedures, training, preventive maintenance, management of change, and periodic audits), and preparing for emergencies. This integrated system must be implemented on an on-going basis. The RMP rule adopts the PSM standard as the basis for the prevention program and streamlines it for facilities that pose lower levels of risk (based on accident history, complexity of the process, and whether a worst case release could affect public receptors).

Many RMP facilities are also covered by PSM, yet continue to experience accidents. OSHA does not require that facilities provide information outside the facility. Because there is no list of facilities subject to OSHA PSM, it is not possible to determine with certainty whether implementation of the PSM standard has reduced the number or severity of chemical accidents for processes subject to the standard. Of the top seven sectors reporting accidents in the RMP however, five (refineries, chemical wholesalers, chemical manufacturers, pulp and paper mills, and cold storage facilities) are almost always subject to PSM for the same chemicals covered by the RMP rule. The other two sectors (drinking water systems, wastewater treatment systems) are covered by PSM in half the states. These sectors account for almost 80 percent of the accidents reported in the RMP history of accidents since 1994. The largest number of releases reported in the RMPs happened in 1997 and 1998 (1999 reports cover only a few months), and the number of accidents reported in the RMPs for these sectors has remained fairly constant over the five-year period. Even if PSM has reduced the accident rate, the number of accidents in these sectors covered by PSM continues to be high.

**CONCLUSION**

Chemical accidents, despite previous regulatory efforts, continue to be a serious problem, causing deaths, injuries, and property damage as well as public and business disruption. Additional efforts are needed in order to improve chemical accident prevention practices.

# CHAPTER 3

# DOES PUBLIC INFORMATION LEAD TO RISK REDUCTION?

A central question for this assessment is "Does disclosure of information to the public lead to risk reduction?" Sharing information that affects the public has long had recognized value in the United States. Congress, in a number of laws, has provided people with access to information that they can use to make better decisions for themselves and society (**see box**). This chapter discusses the evidence that people will use information made available to them and that the result of that use is reduced risk. The next chapter will consider whether the type of information and the ease of access affect the level of use.

## HAVE EPCRA DATA BEEN USED TO REDUCE RISK?

In 1986, Congress passed the Emergency Planning and Community Right-to-Know Act (EPCRA) to improve local planning for chemical emergencies and provide the public with information about hazardous chemicals in their communities. Of relevance to this assessment are sections 312 and 313 of EPCRA.

### Some Public Information Laws and Programs

- Freedom of Information Act
- Federal Advisory Committee Act
- Emergency Planning and Community Right-to-Know Act
- Clean Air Act §114(c)
- Energy Policy and Conservation Act
- Clean Water Act §308(b)
- Resource Conservation and Recovery Act §3007(b)
- Food and Drug Administration processed food labeling program
- Securities Act of 1933, Securities Exchange Act of 1934, and the Securities and Exchange Commission's "EDGAR" database

- Section 312 requires facilities handling more than a threshold quantity (mostly 10,000 pounds) of certain hazardous chemicals to file an annual inventory with the state, Local Emergency Planning Committee (LEPC), and the fire department. The inventory forms are available to the public on request from the state or LEPC. More than 500,000 chemicals are covered by this section.

- Section 313 covers the Toxics Release Inventory (TRI) - annual reporting of releases and transfers of about 650 chemicals. TRI reports are filed with EPA and the states and are made available to the public on the Internet. Before the Internet was widely available, TRI data were available through access to on-line data services (such as the National Library of Medicine). The complete data set could also be obtained.

Section 312 and 313 data differ in some ways from OCA data. Section 312 and 313 data are not interpreted to the extent OCA data are, and section 313 data represent actual releases versus hypothetical releases. Despite these differences, using sections 312 and 313 data for analysis is valid. This use is valid because EPCRA data are collected not only for emergency planning, but also to ensure that the general public (as well as government agencies) is provided with information about chemicals and chemical releases at facilities.

EPCRA section 312 data have not been widely requested or used by the general public. The average number of requests for the data is one per year per LEPC or about 3,500 requests per year nationally (6). Local emergency planners and responders do use the data (e.g., in the 1993 Midwest floods), but the public is generally not aware of the data. In a 1998 report, the General Accounting Office agreed that section 312 data can be useful to a variety of groups, but stated that the use of the data by the wider public has been limited. GAO suggested that costs would be better justified if EPA improved the availability and use of the data by, for example, ensuring that the data are computerized (7).

In contrast, TRI data are maintained in a publicly available database and highlighted in an annual report produced by EPA. From the first publication of the data in 1989, TRI data have drawn considerable attention from community and environmental groups, the press, state and local governments, and industry. Environmental groups have used the data to publish their own reports on specific areas or facilities (**see box**); the press has run numerous articles on local facilities, highlighting those with high levels of emissions. Community groups have used the data as the basis for actions. For example, in Northfield, Minnesota, two citizens groups and the Amalgamated Clothing and Textile Workers Union joined forces and used safety and emissions data to successfully bargain with Sheldahl, Inc. for a phase-out of toxic methylene chloride solvent in favor of a non-toxic substitute, preventing a plant shutdown and loss of jobs. Research shows that TRI data are used by various groups to address environmental risk. Among public interest and environmental groups, the three most frequently reported uses are directly pressuring facilities for change, educating citizens, and lobbying for policy changes. State agencies that run TRI programs report their three largest uses are comparing TRI data to permits, source reduction efforts, and comparing emissions patterns at similar facilities. Industry's two most frequently reported uses are source reduction efforts and educating citizens. Researchers have concluded that TRI data have been used to reduce environmental risk (8). In addition, TRI data have spurred several states, such as Massachusetts and Louisiana, to adopt more stringent state pollution laws (34, 35). **Appendix C** provides summaries of these and a number of other successful uses of TRI and other EPCRA data.

## HAS THE USE OF TRI DATA LED TO RISK REDUCTION?

Since the first publication of TRI data in 1989, the emissions reported under TRI have fallen about 43 percent (9).[7] A number of factors have caused the reduction. In some cases, changes in the methods used to estimate emissions have produced reductions. Environmental groups have sometimes argued that the reductions are mainly the result of these different estimation methods. Industry has argued that the reductions are real, driven by the realization that product was being wasted and opportunities existed to recover valuable materials. Industry has sometimes claimed that their emissions reductions are not stimulated primarily by the publication of embarrassing data or the fear of public attention, yet an informal survey suggests otherwise.

A number of companies reduced their TRI emissions to a far greater extent than the general trend. One of the reasons these companies may have made significant reductions is due to negative press about their environmental performance. This section and **Appendix D** describe an informal assessment whether media criticism and negative press appears to have any positive effect on subsequent toxic emissions (36). This assessment was not intended to be an exhaustive statistical analysis of this hypothesis.

Although a facility may reduce its toxic emissions for a number of reasons, "manufacturers listed among the worst polluters ... may change their ways out of fear of customer boycotts, increased regulation, or community hostility. The company's reputation, hard to build

---

**Some Publications Using TRI Data**

- Toxics Hazards in Los Angeles (California Public Interest Research Group)
- Florida's Toxic Soup: Tracking Toxic Trends
- Toxic Air Pollution in Illinois (Citizens for a Better Environment)
- Clean Water Fund of North Carolina
- West Virginia Scorecard (annual report by the National Institute for Chemical Studies)
- Poisons in our Neighborhoods (state reports, Citizens Fund)
- Poisoning the Great Lakes (Citizens Fund)
- Richmond at Risk (Citizens for a Better Environment)
- Environmental Equity in Louisiana

---

**Impact of Public Information in Indonesia**

Indonesia's Environmental Impact Management Agency implemented a pilot program, known as "PROPER," in which certain industrial facilities were given grades, based on their water pollution performance. Researchers found that public disclosure of these grades was sufficient to prompt 10 factories to invest in pollution abatement in order to improve their rating, and lead to a more than 40 percent pollution reduction in the pilot group in only 18 months (10).

---

[7] Based on 1988 to 1997 TRI data for core chemicals that have been subject to TRI requirements without modification for the entire period.

and easy to destroy, is at stake" (33). Certainly, arbitrary actions to reduce emissions based on fear of regulation or negative press without an understanding of risk can divert resources away from situations that deserve greater attention and public concern. Consequently, it is even more important to ensure that enough information is made public and that the public engage in dialogue at the local level so that attention is focused on real risk reduction.

Several searches of newspapers, trade journals, and magazines revealed seven companies that were repeatedly cited above others as "the worst polluters" in the nation, according to their TRI emissions, and several facilities that were repeatedly cited as "the worst polluters" in their states. (See **Appendix D** for details.) Because of this, the TRI emissions data for each facility were compared before and after it received negative press. In addition, for each of the nine "worst polluting" facilities in states, a comparable facility located in the same region with the same Standard Industry Classification (SIC) codes emitting the same chemicals as those identified as a "worst polluter" was identified. However, these other facilities did not receive negative press for their TRI emissions. Using the TRI database, EPA tracked the toxic releases for the "worst polluting" facilities over the same time period as for the emissions from the comparable facilities.

The seven selected companies that were the subject of nationwide negative press about their total toxic releases reduced their emissions and transfers 1.5 to 2.2 times more than the general TRI trend in toxic releases and 1.3 to 19 times greater than the trends for their specific industry sector as shown in **Table 3**.

**Table 3 – Company-Wide Toxic Release Emissions Reductions 1990-1996**

| Company | Percent Reduction | General Trend | Industry Trend | Improvement Over General Trend | Improvement Over Industry Trend |
|---------|-------------------|---------------|----------------|--------------------------------|---------------------------------|
| Inland Steel | 95% | 43% | 5.2% | 2.2 times | 19 times |
| Kennecott | 90% | 43% | 5.2% | 2.1 | 17 |
| Monsanto | 84% | 43% | 51% | 2.0 | 1.7 |
| American Cyanamid | 83% | 43% | 51% | 1.9 | 1.6 |
| IMC-Agrico | 82% | 43% | 51% | 1.9 | 1.6 |
| DuPont | 73% | 43% | 51% | 1.7 | 1.4 |
| 3M | 65% | 43% | 51% | 1.5 | 1.3 |

At the facility level, the pattern is similar. Each of the facilities that was identified as the subject of repeated negative local press reduced their emissions and transfers following the negative press coverage. The comparable facilities did not. In one case, the facility reduced its emissions of the chemical that was the subject of press coverage by 86 percent and its total facility emissions by 64 percent. In contrast, all emissions and transfers for the other facilities owned by the company stayed relatively the same. (See **Appendix D** for more details.)

**Appendix C** presents additional evidence that companies respond to negative publicity and public pressure by reducing emissions and, therefore, reduce the risk to the public and the environment. Other factors certainly also produce emission reductions, but facilities subject to public pressure do appear to respond to that pressure and reduce the risk that is subject to reporting. This conclusion is further supported by the states that have had chemical accident prevention programs in place under state law since the late 1980s and early 1990s. In Delaware and Nevada, a number of facilities that were complying with the state chemical accident prevention laws switched chemicals or reduced inventories to avoid having to file an RMP with EPA that would become publicly available.[8] In both states, reports filed with the states are available only on request[9] (14, 15).

> **Publicity Brings Risk Reduction**
>
> The power of press attention to produce action was seen in Washington, D.C. in late 1999, when the *Washington Post* ran a series of articles on concerns for chemical safety at the Blue Plains Wastewater Treatment Plant. The articles featured scenario information that the plant (voluntarily) included in its RMP Executive Summary, the health effects of the chlorine and sulfur dioxide used at the plant, and apparent problems with plant safeguards. As a result of these articles, the **D.C. Water and Sewer Authority** and the **mayor** immediately committed to resolving all of the safety problems cited in the articles. The Authority has improved security and now plans to replace its chlorine one and a half years sooner than originally planned (11)(12)(13).

Note also that with the growth in the economy, chemical use has increased. Industrial production between 1991 and 1997 (the last year of TRI data) increased by 28.4 percent. Chemical industry production increased by 19.5 percent. Consequently, the 43 percent average reduction in TRI emissions and transfers reflects an even greater reduction per unit of production.

## HAS THE PUBLIC FOCUS ON HAZARDOUS CHEMICALS REDUCED ACCIDENTS?

Accidental releases can be more difficult to prevent than routine releases (TRI covers both) because the causes of accidents vary considerably; serious accidents are often the result of a series of failures (human and equipment). A decline in the number or severity of accidental releases would indicate that facilities have improved practices generally.

---

[8] Other explanations for why these facilities exited the Delaware and Nevada programs are unlikely. For example, the sources that exited these programs are likely to have already incurred most of their compliance costs due to the similarity of EPA's and the state's requirements (other than disclosure).

[9] New Jersey and California also have state programs. More than two thirds of the original registrants in New Jersey switched chemicals or reduced inventories to avoid the state law, which involves large annual fees ($10,000 to $100,000). It is unlikely that those that remain in the program have the option of avoiding the federal program. California delegated its program to 147 local agencies and keeps no central records so that trends in that state cannot be traced. California also uses much lower chemical thresholds than the federal program.

EPA examined whether certain hazardous substance releases have declined in the 1990's. As noted in **Chapter 2**, facilities are required to report episodic releases of hazardous substances that are above a reportable quantity (RQ) to the federal government. Data from these reports are compiled in EPA's Emergency Response Notification System (ERNS) database. ERNS also includes petroleum product spills (crude oil, processed oil, diesel, jet fuel, kerosene, gasoline); the database covers both transportation and fixed facilities. From a high of 7,800 in 1994, hazardous substance releases recorded in the ERNS database declined to 5,400 in 1999.

Because ERNS data include many reports of unknown material or unknown sources, and for hazardous substances, many reports of substances that are not subject to reporting and of quantities below the reportable quantities, EPA narrowed the analysis to four states, Massachusetts (MA), Connecticut (CT), Virginia (VA), and New Jersey (NJ). Focusing on four states allowed for a more detailed analysis of the data. These were selected because the four were relatively similar in size and industrial sectors and include a broad range of industries. The analysis looked at all ERNS-reportable hazardous substance releases from fixed facilities, excluding those generated by private citizens or unknown sources and those of unknown materials. The analysis also looked at TRI chemicals reported by manufacturers, and finally, all reportable releases based on the current RQs for currently listed substances. Because the reportable quantities and chemicals listed changed in the early to mid 1990s, EPA made sure to use consistent criteria when comparing reported releases.

The results of all the analyses were similar, as can be seen in **Exhibit 1**. In these four states, all hazardous substance releases in ERNS from fixed facilities declined by 60 percent from their peak year (1992). Similarly, episodic releases of TRI chemicals from manufacturers (the only group subject to TRI) and releases of hazardous substances above current RQs declined by about 68 percent from their peak year (1990). Because the analysis considered only currently listed substances at current RQs, the 1995 RQ adjustments are not responsible for the decline. (Note that these numbers cannot be compared to national hazardous substance numbers, which include a substantial number of releases from unknown sources and of unknown materials, as well as some releases generated by private citizens.) **Appendix E** provides the data from the four states.

**Exhibit 1 – Episodic Hazardous Substance Releases – Four States**

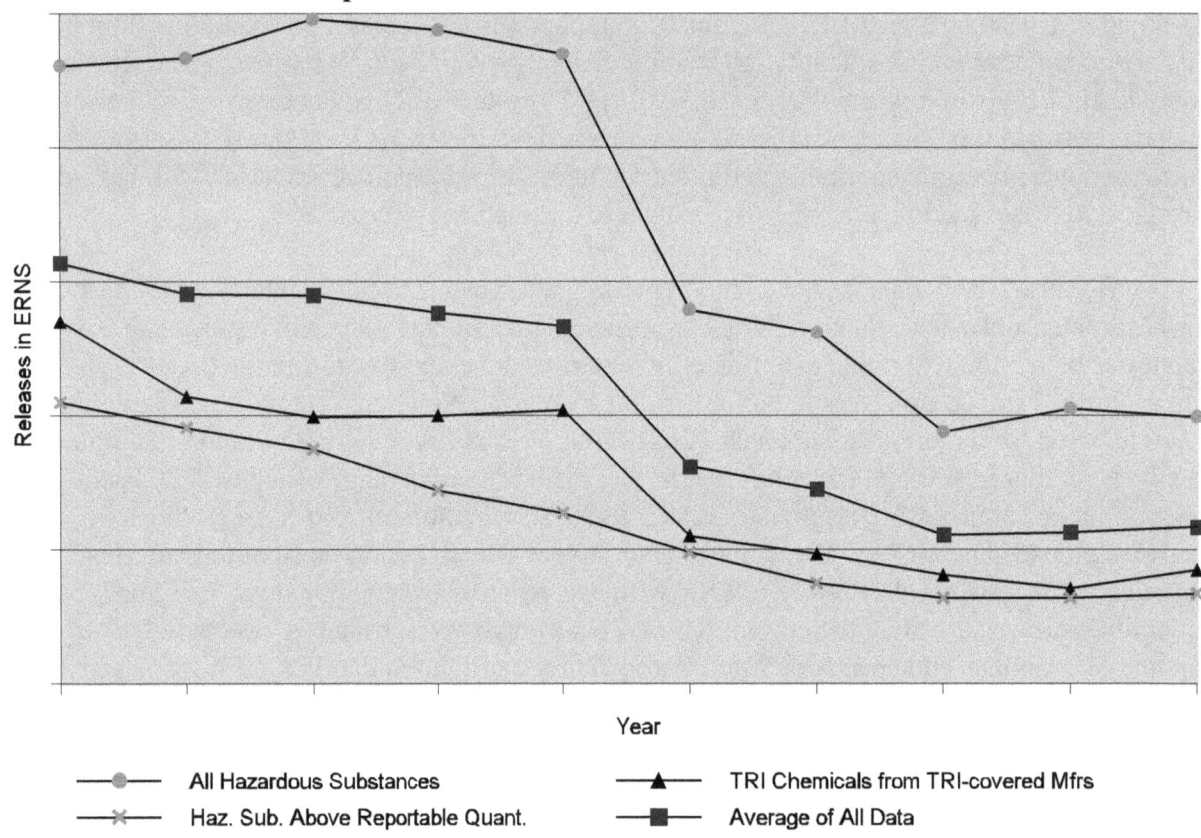

The analysis also looked at petroleum product spills and transportation releases in the four states. In contrast to hazardous substances released from fixed facilities, petroleum product spills and transportation releases are not subject to extensive right-to-know disclosure. Although oil and gasoline are subject to reporting under EPCRA section 312, they are not covered by TRI (refineries are only covered for toxic chemicals in petroleum products and chemicals they use to process oil). Non-petroleum-based oils (e.g., vegetable oil) were not counted. As can be seen in **Exhibit 2**, there is no visible trend in reportable spills. The number of spills varies from year to year, without a consistent pattern. Nationally oil and transportation spills have declined, but show the same pattern of variation up and down from year to year.

All else being equal, hazardous substance releases and oil spills from fixed facilities and transportation would be expected to rise. As noted above, manufacturing production has increased 28 percent since 1991. Production of petroleum products has increased 18 percent; the Department of Transportation estimates that the number of delivery trips increases by about two percent a year. Without improvement in safety practices, the number of all types of releases would be expected to rise. In addition, it may be that increases in the early part of the decade reflect improved reporting compliance for all types of releases. The reduction in the number of reportable hazardous substance releases, therefore, indicates that facilities that use these substances, subject to public scrutiny to a far greater extent than oil and transportation, have improved their handling substantially and reduced the risk to the public to a greater degree than

22

**Exhibit 2 – Oil and Transportation Releases – Four States**

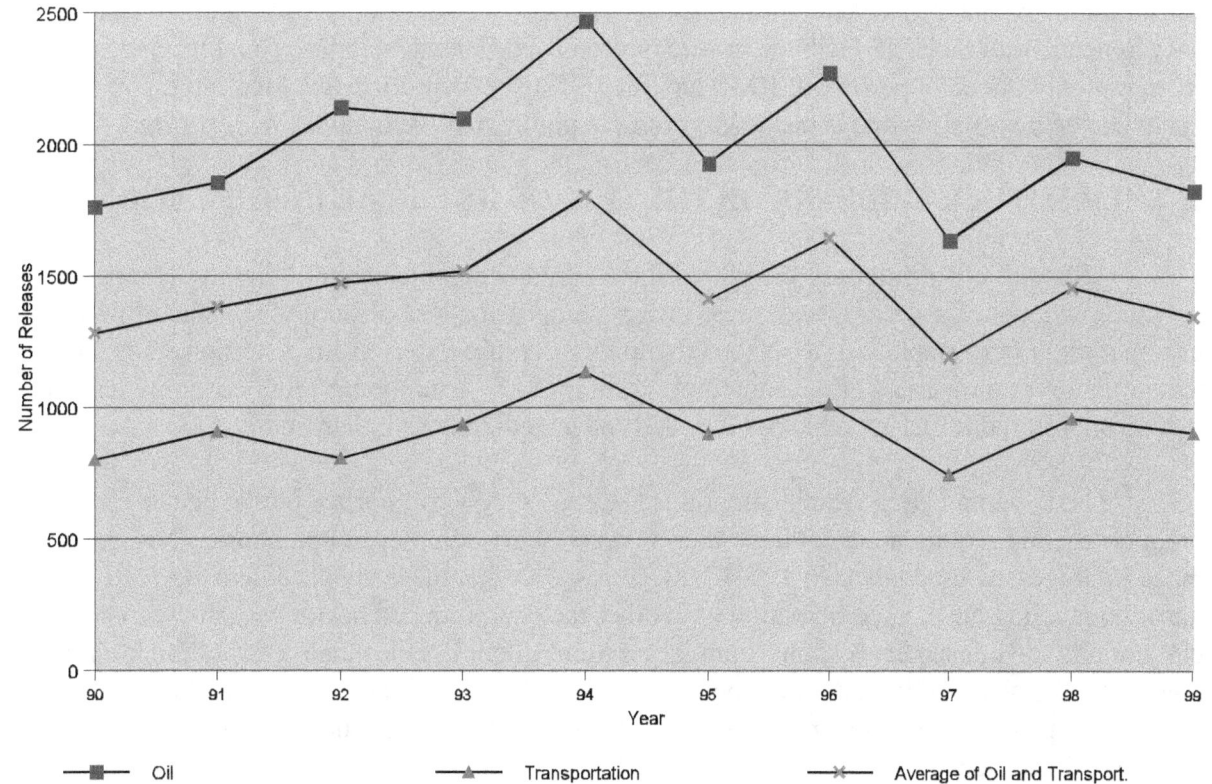

has occurred for substances and modes not subject to the same level of public scrutiny.

## IS THERE EVIDENCE THAT OCA INFORMATION AVAILABILITY WILL LEAD TO PUBLIC ACTION?

Because RMPs were filed just last year and the public has had little or no access to OCA information, there has been little opportunity for the public to use RMP or OCA data to pressure industry to reduce risks. The chemical industry, however, has conducted pilot projects to provide the public with OCA data. In 1994 in the Kanawha Valley of West Virginia, industry, in conjunction with local emergency planners, developed worst-case release scenario data and held a public meeting to inform citizens about the risks posed by hazardous chemicals in their community. The event was held in a local shopping center that citizens frequented on Saturdays. Booths were set up and information was distributed that explained the chemicals used by local industries and what health effects could occur if those chemicals were accidentally released. Armed with this information, citizens since have formed groups and have gone to companies to persuade them to discuss their overall environmental management strategy, including using less toxic substances. Additionally, other citizens worked with industry officials to prepare and distribute a kit to community residents that included instructions and materials that could be used to shelter-in-place during a chemical emergency (16).

Such pilot projects have been led by the chemical industry, which has made outreach a part of its Responsible Care® program. These voluntary efforts, though laudable, are rare, and have generally not included facilities in other sectors. The chemical industry and refineries represent only about 13 percent of RMP facilities. Thus, except for a few locales, the outreach about accident scenarios that is needed to promote public involvement has not occurred.

## IS THERE OTHER EVIDENCE ON THE IMPACT OF PUBLIC INFORMATION?

Although other programs outside of the environmental area are not directly comparable to the OCA situation, the success with these programs none-the-less illustrate the value of public information. For example, a 1981 National Highway Traffic Safety Administration study found that, after 1976, purchasers of new automobiles in the U.S. actually used fuel economy test data to help choose fuel-efficient vehicles (17). A 1985 study made a similar finding (18).

The Food and Drug Administration (FDA) has sought to improve public health via product labeling requirements for a number of years. For example, comprehensive changes were made to processed food labels in 1994. Research shows that these enhancements to food nutrition labeling have, in combination with other factors, produced health benefits. Studies show that not only do consumers read nutrition labels more often, but they also reduced the fat in their diets (19). Now, based in part on the positive effects of the 1994 label revisions, the FDA has recently proposed adding label requirements for trans-fatty acids, to further reduce rates of heart disease and early death.

There are many ways in which access to OCA information could yield positive impacts, based on incentives created or affected among various segments of the public. These incentives and the likely uses of OCA information by the public are detailed in **Appendix F**.

## CONCLUSION

The public (which includes the media, public interest groups, as well as citizens) does use available information on risks and the availability of such information correlates with a substantial reduction in the risks that are the subject of reporting.

# CHAPTER 4

# DOES THE TYPE OF INFORMATION AND TYPE OF ACCESS MAKE A DIFFERENCE?

The first part of this chapter addresses two questions: Does the type of information available influence the degree of risk reduction caused by its availability? Does the type of access to the information affect the use of the data?

## A. DOES THE TYPE OF INFORMATION INFLUENCE DATA USE?

RMP data other than OCA data are already available on the Internet through RMP*Info[10]. The RMP data contain information that could be used to generate off-site consequence results, although these would be less accurate than the OCA data reported by facilities (see **Chapter 5**). OCA data are potentially valuable because they are "interpreted" data. For example, rather than simply providing the public with raw numbers for the quantities of chemicals on-site, the OCA data provide the public with information about what might be affected if a worst-case or alternative release involving that quantity should occur. The public does not need to do anything further with the data to understand why it should be concerned (or not) about chemicals at the facility.

TRI data are partially interpreted data. They tell the public how much of a toxic chemical is being released into the community and the media (air, water, etc.) to which it is released. But unless the public is familiar with the hazards of the chemical and industry practice for handling that chemical, the public may not understand whether the emissions pose a risk and whether those risks are high or low in comparison to other facilities in a locality or across the U.S. TRI data have always been available electronically. The data were originally posted in the on-line database of the National Library of Medicine and could be searched both by facility and on combinations of other parameters. TRI data are now available in at least three Internet databases:

- EPA's Envirofacts database (www.epa.gov/enviro), which presents some of the data with simple queries; more advanced queries are possible, but not simple to program. No interpretation is provided.

- RTK-Net, which provides access to all data on each TRI form, without interpretation. RTK-Net (www.rtknet.org) also provides access to multiple other environmental databases.

---

[10] RMP*Info™ is the name of EPA's database of RMPs submitted by facilities. RMP*Info™ is located in the Envirofacts Warehouse on the Internet at http://www.epa.gov:9966/srmpdcd/owa/overview$.startup.

- Scorecard, the database from Environmental Defense (formerly Environmental Defense Fund), which was designed for easy use and provides interpretations of TRI and other environmental data on each facility. For each facility, Scorecard (www.scorecard.org) provides not only chemical names, but the hazards of the chemicals, how the facility ranks in the U.S., state, and county on a number of measures (e.g., total releases, cancer risk, non-cancer risk), quantity of releases aggregated by hazard, estimated cancer risk, and the facility release history for each year since 1987. The site also allows a user to map the facility in the county and to send a free fax to the facility.

Public reaction to the Internet launch of Scorecard in April 1998 is a measure of the value of interpreted data. At that time, RTK-Net was registering about 240,000 searches of its databases a year. Scorecard has registered 2 million visitors since its launch and serves about 600,000 page views[11] *a month*. Site visitors have sent about 5,000 faxes to about 3,000 distinct companies (20). By providing more highly interpreted data, Scorecard has increased the accessing of data on the Internet by a factor of up to 30. As a point of comparison, from July 1999 through January 2000, without OCA information, EPA's RMP*Info has logged over 155,000 page views of RMPs or about 23,000 per month (or about four percent of Scorecard's page views). A comparison with Scorecard is not entirely fair; Scorecard provides other environmental data and ease of use, while RMP*Info is new, concerns only accidental releases, and does not contain OCA information.

Another reflection of the value of interpreted data is that in the early years of TRI, few reporters used the TRI database directly (only 20 percent in one review of large dailies); the large majority waited until the government or public interest groups published interpreted data (21). Overall, evidence strongly supports the argument that the form of the data does affect the degree to which the public will use it. Interpreted data, such as OCA data, will be used far more than raw data.

## DOES EASE OF ACCESS TO THE DATA AFFECT USE?

As mentioned in the previous chapter, EPCRA section 312 requires thousands of facilities to file annual forms reporting on the quantities of hazardous chemicals they use or store. Section 312 covers all hazardous chemicals (estimated to number about 500,000) and almost all industrial sectors. The 312 forms, therefore, may be the most complete record of hazardous chemical use in the U.S.

These forms are filed with local emergency planning committees (LEPCs), state emergency response commissions (SERCs), and local fire departments. LEPCs and SERCs are required to make the forms available to the public, but, because of limited resources, only about

---

[11] A "page view" is the downloading of an HTML page. The number of Scorecard page views would tend to be somewhat higher than the number of database searches for the same activity, because most but not all of its page views pull data from a database. "Hits" are not used here because they are less accurate indicators of data access.

**Exhibit 3 – Impact of Access and Type of Data**

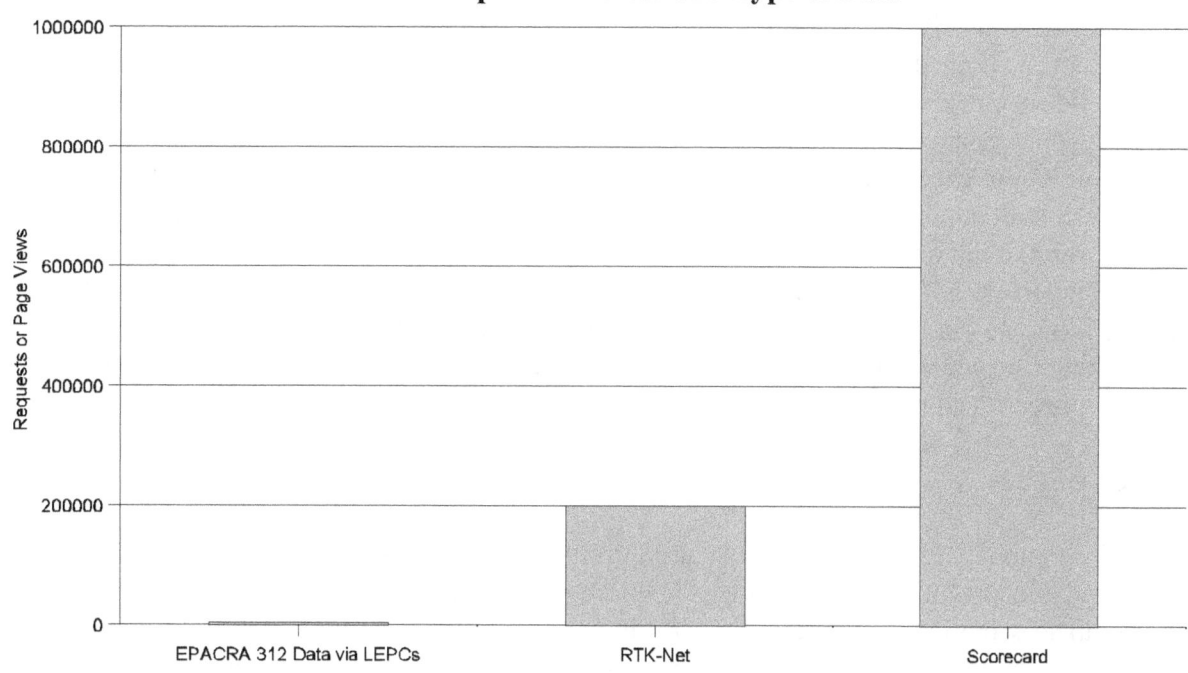

half advertise the forms' availability. If a member of the public wishes to see the forms, he or she must locate the SERC or LEPC and request the information. According to a survey conducted for EPA by George Washington University, the average number of requests for these forms is one per LEPC per year (22). The national total, therefore, is about 3,500 requests per year. In short, obtaining section 312 data requires an effort to seek it out; few people have made the effort. A further disincentive, as noted above, is that the data are often not laid out in a format from which the public can easily gain value.

The results are similar for state chemical accident prevention programs. As discussed in **Chapter 3**, New Jersey, Delaware, and Nevada have had accident prevention program rules in place since the late 1980s and early 1990s. In each state, facilities file reports, but these are not made available to the public except on request. New Jersey reports that since its program began in 1988, less than 10 individuals or organizations have requested facility reports or information (23). Delaware has received similarly few requests (24). Nevada places its reports in public libraries, but its experience indicates that librarians have to hunt for the files when updates are needed. Few people have requested information from the state unless there has been an accident(25).

This lack of interest and effort contrasts sharply with the number of people seeking data from RTK-Net and Scorecard, where the data can be accessed at any time from office or home. Comparing the 3,500 requests a year via LEPCs to the 240,000 database searches per year for Internet-based RTK-Net is consistent with the finding that providing data on the Internet has increased the use of data many times over. Comparing the LEPC request number to Scorecard visitors per year yields a difference of a factor of more than 280.

CSISSFRRA requires facilities to hold a public meeting or post a notice summarizing their OCA information. EPA has not surveyed facilities to determine attendance at these meetings; however, a very limited number of facilities have provided feedback. One large California public agency held seven meetings in different parts of Los Angeles, Orange, and Riverside counties. The average attendance was two to five, though at the final meeting, held on a Friday night, one person came. For each meeting the agency ran an ad twice in the local paper and sent 500 to 1,000 direct mail announcements to households within one mile of the facility. Other reports have been similar, though one meeting covering several facilities had 500 people attend. That meeting was in Henderson, NV, the site of two major accidents in the last 11 years. In contrast, when industry has gone out to the public (e.g., at a shopping mall in Charleston, West Virginia) rather than making the public come to industry, outreach and communication about chemical accident risks have been much more successful.

**CONCLUSION**

The public will make use of data and information to the extent the information is understandable, useful, and easy to use. The more effort is required on their part, the less likely they are to seek out the data. If data are easy to access, use will rise significantly. If data are interpreted rather than raw, use will expand even more. **Exhibit 3** shows the numbers of requesters for section 312 data sought from LEPCs, raw data from RTK-Net, and interpreted data on Scorecard.

# CHAPTER 5

# ARE THERE OTHER SOURCES OF THE SAME DATA?

Another question that must be considered is whether the public already has access to OCA data or can easily derive such data without the dissemination of the OCA sections of RMPs. This chapter considers what is currently available and whether this information is an adequate substitute for the data submitted as part of the RMP.

## WHAT OCA OR OCA-RELATED DATA ARE AVAILABLE TO THE PUBLIC NOW?

One potential source of OCA data is the RMP Executive Summary. By regulation, each RMP's Executive Summary must include a brief description of the worst-case release scenario(s) and the alternative release scenarios(s), including administrative controls and mitigation measures. These Executive Summaries are posted on the Internet. However, neither regulations nor guidelines from EPA specify how much information should be in the Executive Summaries. They vary greatly in the amount of OCA data they provide. In addition, they are in narrative, or text, format, so they cannot be easily searched on variables of interest. Executive Summaries may or may not provide information of interest regarding a particular facility, and they cannot be relied upon to provide a full picture of a group of facilities or a basis for comparing facilities.

Other publicly available data can be gathered and analyzed to provide information similar to some of the OCA data. For example, important components of the worst-case release scenario are the identity and the quantity of toxic or flammable substance held in the largest vessel at the site. The chemical name and quantity of that chemical in each covered process is available in the registration section of the RMP. However, for a large facility with several listed substances or covered processes, the public will not know which chemical or process was considered for the potential worst-case or alternative release or how much of the chemical was expected to be released. Even if only a single chemical is at the site, the quantity of the chemical in the process may not be the same as the quantity stored in the largest vessel.[12] Storage quantity is also available through TRI and other Internet or publicly available databases, but these data reflect total quantity on site, reported in broad ranges. If the quantity of the regulated substance listed in the RMP or provided by TRI is used, it is likely that the scenario would over-estimate the consequence distance since the maximum quantity in a process or on-site is being used in lieu of the maximum amount stored in a single vessel.

For the worst-case scenario, the public has access to an EPA model or software tool which most facilities have used to perform their analyses; however, without certain key inputs (quantity in the vessel and specific chemical), people may get results that are very different (and

---

[12] In some cases, the largest vessel may not be the source of the worst-case release if a release from another vessel closer to the public would have greater off-site impacts.

potentially more alarming) than those provided by the facility. For alternative scenarios, the facility has wide latitude in its choice of inputs; without these same inputs, a member of the public has no way to reproduce a facility's OCA data.

The potentially affected population is a function of the distance that a toxic cloud travels or that an explosion affects, so whatever errors are made in determining this distance will also affect the public's calculation of population affected.

In contrast to information dependent on the release distance, more general siting and surrounding population data is more easily accessible. Census and other data for determining the population of a given area are publicly available. Facility location information can easily be found in telephone directories (e.g., yellow pages). In particular, sources of the yellow pages on the Internet are also linked to a map that shows the exact location of facilities. Other siting information for major facilities of publicly traded companies appear in 10-K reports available on the Internet through the Securities and Exchange Commission's Electronic Data Gathering, Analysis, and Retrieval (EDGAR) system.

The prevention program portion of the RMP lists all the types of mitigation measures used by a facility in a process, but without OCA-specific data the public will not know which measures should be involved in calculation of consequence distance or the extent to which the measure might mitigate a release.

## HOW EASY WOULD IT BE TO REPLICATE AN OFF-SITE CONSEQUENCE ANALYSIS?

Data are available to conduct an OCA, even if the results are likely to differ considerably from those submitted by facilities. How easy is this process for a citizen?

Accessing and reading the regulations and the guidance would be an initial step, followed by acquiring the mathematical models needed to turn quantity data from the RMP or other sources into concentration estimates around the facility. Information on the locality around the facility (population, land-forms, buildings and other structures, environment, meteorology, etc.) would need to be acquired. Information peculiar to the facilities would need to be gleaned from the raw data in the RMP, if, indeed it was actually available. Finally, the user would have to integrate this information and tools, check the results, and produce estimates. Although members of the public are not likely to perform such tasks, determined individuals or organization(s) could do so relatively easily using information readily available.

Clearly additional economic costs to society for producing such estimates by outside parties would be incurred. The level of costs would depend on how many such independent assessments might be made, and how many of these assessments might be made by different parties; it would be impossible to estimate with any certainty. However, it seems very likely that the summation of the costs of outside assessments would be considerably greater than the costs that have already been incurred by the facilities in producing one OCA per facility. These are sunk costs, (i.e., already incurred), and were estimated at the time of the RMP rulemaking at $44 million, nationwide. OCA-specific information already exists. Any re-estimation by outside parties would entail additional societal costs beyond those which have already been expended. From the point of view of economic efficiency, having the OCA analysis produced once (by the facilities) is far preferable to having outside users attempt to do so themselves.

## DOES THE OCA INFORMATION PREDICT ACTUAL RELEASES OR ACTUAL CONSEQUENCES?

As described in more detail in **Appendix A**, the OCA information represents hypothetical estimates of the consequences of worst-case (WCS) and alternative release scenarios (ARS). They do not predict likely releases or consequences. Worst-case scenarios in particular are based on extreme conditions. Many of these conditions cannot be controlled or predicted. Some of these conditions and a description of their characteristics are:

- Weather. The WCS is evaluated at very low wind speeds and very stable atmospheric conditions like what might occur at dawn on a warm summer morning. A smoke plume from a fire-place might look like a very lazy ribbon as it travels slowly downwind (due to less dispersion). This condition occurs fairly infrequently (26); most of the time there is a slight breeze and some instability in the atmosphere which acts to quickly disperse a toxic cloud. A worse condition that sometimes occurs, especially in the summer, is an atmospheric inversion which can magnify pollution effects. The ARS is evaluated at more "typical" weather conditions - moderate winds and atmospheric stability. No one can control the weather conditions at the time of an accidental release.

- Affected population and wind direction. The OCA information contains estimates of affected populations inside worst-case and alternative release scenario circles whose radius is equal to the downwind distance the cloud travels to a toxic endpoint (see below). Toxic gas clouds generally travel in the direction of the prevailing wind; they form a long, narrow plume, which covers only a relatively small fraction of the worst-case circle. (See **Exhibit 4.**) Therefore, depending on wind direction, only a small fraction of the people reported in the OCA information would actually be affected by a cloud. In a sense, the population reported is a sum of the people under all worst-case plumes emanating from a location; any individual plume will cover far less than the sum. For flammable gas scenarios, the blast effects would be felt in all directions from the source, so all people inside the circle could feel the effects, depending on the characteristics of the blast.

**Exhibit 4 – Typical Plume Map (27)**

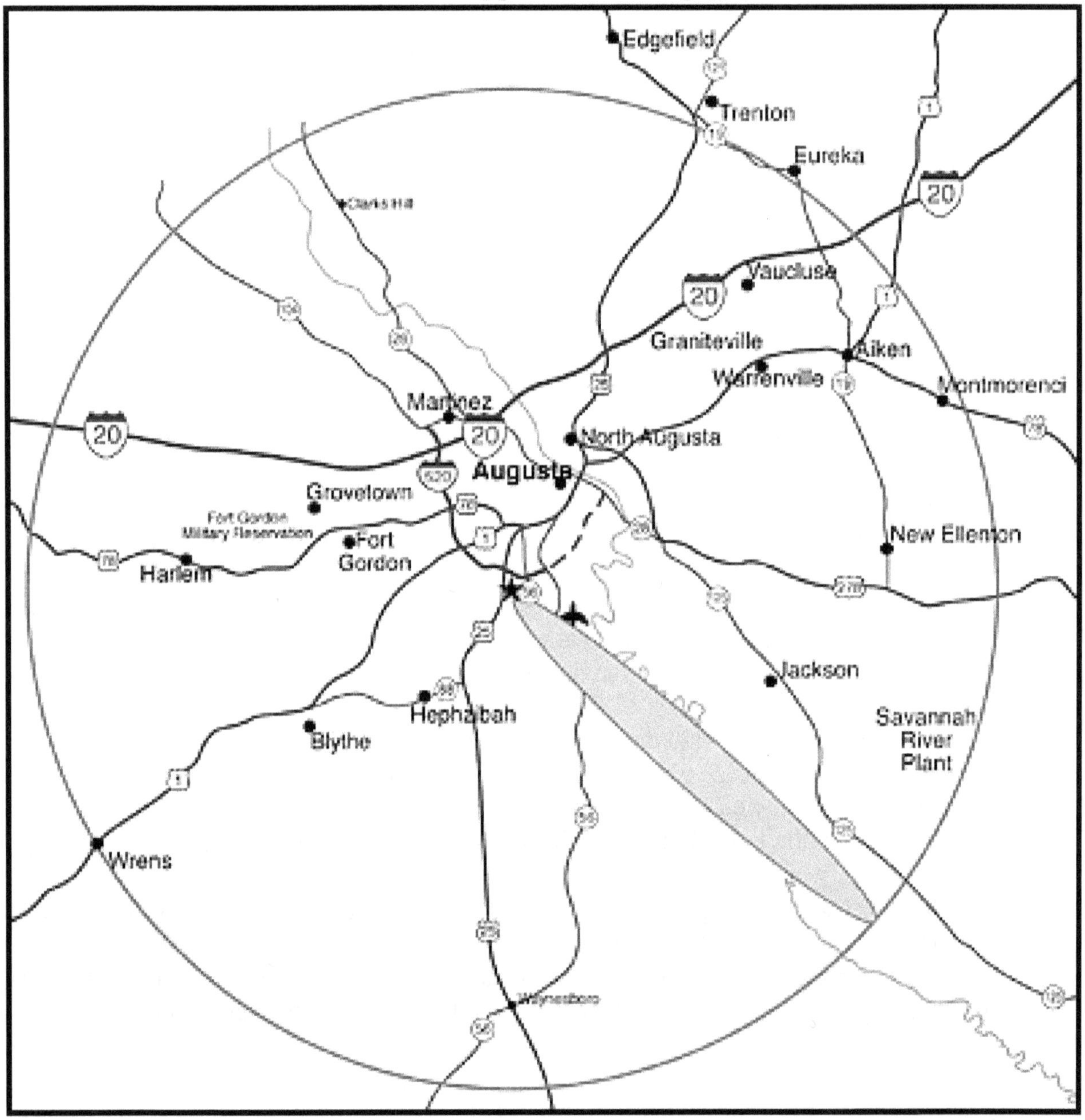

- Actual Exposure or Endpoint. As mentioned above, the OCA information includes an estimate of the population affected by a WCS or ARS within a circle whose radius is determined by the distance from a point of the release to a toxic or flammable endpoint. This number is *not* an estimate of the number of fatalities or injuries that would occur following an actual release, because the endpoint is not the concentration where there is a

100% likelihood that an exposed person would die or suffer injuries. The concentration that would result in death, for example, is much greater and would be much closer to the source of the release. The population number represents the people that could potentially be exposed; actual exposure and the effects of exposure would depend on where a person is (indoors or outdoors) and for how long he or she remains in the gas cloud, and breathing the chemical. For flammable substances the endpoint for a worst-case explosion is only one pound per square inch (psi) over-pressure; fatality levels are much higher (ear-drum rupture is 2.3 psi).

- Amount Released. The WCS assumes immediate (or within 10 minutes) release of the maximum quantity of chemical in the largest storage vessel. However, there is no guarantee that a vessel will always hold the maximum amount; this may occur infrequently. In addition, chemicals flow out or vaporize from failed vessels according to the laws of physics; there may be an initial surge followed by a gradual reduction in flow, taking more time than that given by the WCS. Modeling these conditions is complicated; the WCS is designed to be a simple, straight-forward estimate.

Studies of severe chemical plant accidents have shown that such accidents have usually resulted from the confluence of multiple abnormal events or conditions in process or management systems along with unusual meteorological conditions over which no one has control (28). For example, in Bhopal, four separate safety systems, any one of which would have prevented the accident, had been disabled prior to the accident, and a fifth failed to operate properly (29). Further, weather conditions were such that the chemical cloud did not disperse, exposing more people than if it had dispersed.

In the case of flammable materials, an initiating explosion will most often immediately ignite any flammable material released, causing a large fire, but possibly preventing a much more severe vapor cloud explosion (the worst-case flammable accident). Vapor cloud explosions require that flammable gas be released into a somewhat confined area for a period of time prior to gas ignition. If the gas is immediately ignited before sufficiently large quantity is released, a fire usually occurs, but generally not an explosion. Numerous experimental programs devoted to the study of vapor cloud explosions have shown that such explosions are difficult to reproduce, even under carefully controlled conditions (30) . Even accidental explosions at facilities that store, transport, and manufacture explosives have usually resulted in little off-site damage (31).

## CONCLUSION

Although the public could educate itself and attempt to reproduce OCA data from other sources, the effort involved is likely to discourage the vast majority of people. At the same time, some OCA data are available in RMP Executive Summaries, which are available on the Internet.

Actual chemical releases are different from the releases evaluated in the OCA information. No one can control all of the conditions (for example, weather) used to develop the OCA information; actual conditions at a facility can vary widely from those used in the OCA assessment.   However, facilities subject to the RMP requirements generate the OCA information according to consistent guidelines so that the public and others can understand the relative potential hazards and risks present at a facility.

# CHAPTER 6

# HOW MUCH INFORMATION IS NECESSARY TO SPARK RISK REDUCTION EFFORTS?

The previous chapters describe how OCA data might be used by a variety of stakeholders for chemical accident risk reduction. In order for OCA data to be most effective as an incentive for risk reduction, data from a sufficient number of facilities must be disclosed to the public. Disclosure of a sufficient number of facilities' OCAs is necessary to identify plausible relationships between the hazards posed by facilities and their accident prevention and mitigation actions and the risks of future chemical accidents. A plausible relationship can be determined from comparisons between similar facilities (or processes) and chemicals. This chapter will provide several options for the public to obtain comparison information based on OCA data as well as the number of RMPs submitted that would support each option.

## WHAT INFORMATION IS IMPORTANT FOR RISK REDUCTION?

Risk is a measure of the potential for economic, environmental, or human loss. Chemical accident risks are generally estimated as a product of the quantity and toxic or flammable characteristics (chemical hazards) of a substance, the population potentially exposed by the release, and the likelihood that the chemical will be accidentally released. For example, if a small release of hazardous chemical occurs frequently (indicating that it is likely), but the chemical does not generally migrate off-site, the overall risk to the public is probably low. If the likelihood of a catastrophic release of a large quantity is extremely low, but the number of people affected if it did occur is large, the overall risk may still be low because of the low probability of release. If a large release could occur relatively frequently affecting a large number of people, the overall risk to the public is high.

All else being equal, risk reduction occurs when either the chemical hazard, population potentially exposed, or likelihood are reduced. The toxic or flammable characteristics of a chemical generally cannot be changed; for example, chlorine is toxic when inhaled, propane is flammable. However, reducing the quantity of hazardous chemicals on-site or switching to less hazardous substances will reduce risk. The population potentially exposed can be reduced by minimizing the amount available to be accidentally released or by mitigation measures that capture or destroy an accidentally released substance before it travels off-site. A relative sense of the likelihood for accidental releases at a particular facility can be gained from its accident history; reducing the likelihood of accidental releases is the main goal of the RMP program.

RMP and OCA data give the public some insight into all of these factors. However, the most straightforward ways the public is likely to measure whether risk is being reduced are a reduction in the number and severity of accidental releases and population potentially exposed. The distance to an endpoint and the estimated population and public receptors within this distance serve as indicators of the potential size and impact of an accidental release at a facility.

## HOW IS INFORMATION USED TO ENCOURAGE ADDITIONAL RISK REDUCTION?

Risk reduction begins when an individual who can initiate actions for risk reduction (e.g., facility management, a community member, local media) believes that the current level of risk is unacceptable. Dialogue at the local level among the public, the LEPC, first responders, and the facility is the best way to raise concerns and to discuss possible options.

RMP and OCA information from a single facility provide insight about risk that could affect an individual. But by itself, developing a clear understanding of risk and whether it is acceptable is difficult. Since everything we do is "risky" to some extent, risk-based decisions must rely on a comparison with other familiar risks or known benchmarks; consequently a sense of the risk posed by one facility is best established by comparing it to other facilities. From this comparison and understanding of potential risk, unacceptable risks can be reduced through a reduction in hazards (e.g., lower quantities of chemicals on site) or likelihood (e.g., improving accident prevention programs or mitigation systems) or both.

One simple approach to comparing the hazards posed by facilities is to compare distances to an endpoint. An individual can compare the distances for multiple facilities and obtain a relative ranking of the maximum and more likely extent of risk posed by the facility. Distances to an endpoint can be compared because the distance to endpoint represents a similar level of potential impact. Likewise, the residential population and the public and environmental receptors within the distance to endpoint provide additional insight into the potential extent of the damage associated with a release and, hence, an aspect of risk. However, two facilities could have the same endpoint distance but at one facility no public receptors are within this zone. These approaches, however, do not convey information about the likelihood of a release. For example, given two facilities with the same distance to endpoint or similar number of public receptors, one could have a much higher likelihood of a release, and thus a higher risk.

## HOW MANY RMPs OR OCAs ARE NECESSARY FOR FACILITY COMPARISONS?

### A. Comparisons across facilities in the same locality

An individual may want to compare the distances to an endpoint for a particular facility with distances for all other facilities within a given locality, for example, a county, or, if close to a county border, two or more counties. (Most LEPCs are county-based, and a number of facilities have distances to endpoint of 25 miles.) **Table 4** shows the counties with the most RMP facilities. Individuals within these counties would need access to at least this number of complete RMPs (including OCA information) to develop county-wide comparisons. Alternatively, an individual could collect the RMPs within a state, city, or zip code; but because of wide variability in the number of facilities within states, cities, or zip codes and since most LEPCs are established by county, a county-wide comparison seems to make the most sense.

**Table 4 – Counties with the Most Facilities Submitting an RMP**

| County: | Number of RMP Facilities: |
|---|---|
| Harris County, TX | 201 |
| Los Angeles County, CA | 162 |
| Kern County, CA | 96 |
| Cook County, IL | 76 |
| Maricopa County, AZ | 71 |

**Table 5** shows how the number of RMP submitting facilities are distributed across counties. The RMP database has RMPs from 1,501 to 2,384 counties in the US; these counties have at least one RMP up to 201 RMP facilities (Harris County, TX). Ideally, residents should be able to see all the RMPs and OCA information for the facilities in their county; if they are restricted to 10 RMPs for example, residents in 82% of the counties that have RMP facilities would be able to review all RMPs in their county while 18% would be unable to do so.

**Table 5 – Distribution of RMPs and Counties**

| Number of RMP Facilities | Number of Counties | Portion of all Counties that have RMP Facilities (percent) |
|---|---|---|
| 1 - 5 | 1501 | 63% |
| 1 - 10 | 1955 | 82% |
| 1 - 15 | 2222 | 93% |
| 1 - 20 | 2309 | 97% |
| 1 - 50 | 2376 | 99% |
| 1 - 100 | 2383 | 99% |
| 1 - 201 | 2384 | 100% |

But comparing facilities based on distance to an endpoint within a county is not likely to be enough. The public needs to know more to be able to compare facilities and form an understanding of the risk. For example:

- Is the quantity of chemical stored reasonable?
- Are there mitigation or other measures that could be used to reduce the distance to endpoint?

To answer these questions, the public needs to be able to compare RMPs for similar facilities using the same chemicals.

## B. Comparisons based upon Type of Process

The best way to compare a process at one facility with other similar processes at other facilities is to choose processes that are categorized in the same industry sector. When a facility prepares an RMP, processes are assigned a code using the North American Industry Classification System (NAICS). The NAICS codes identify the industry segment for that process. For example, Agricultural Supply processes have a NAICS code of 42291. If someone wants to compare a process with all others in its industry sector, then all RMPs that have a process in the same industry sector would be needed.[13] **Table 6** lists the top twelve industry sectors (based on NAICS code) that had the greatest number of submitted RMP processes. For example, if someone wants to compare OCA data from an agricultural supply process (NAICS 42291) in their area to all other agricultural supply processes, then the OCA data from 4,033 other agricultural supply RMP processes would be needed.

**Table 6 – Industry Sectors Submitting the Most RMPs**

| Industry Sector Description - NAICS Code | RMP Processes | Percent of All RMP Processes |
|---|---|---|
| Agricultural supply - 42291 | 4,034 | 27 |
| Water Treatment - 22131 | 1,892 | 13 |
| Sewage Treatment - 22132 | 1,361 | 9 |
| Refrigerated Warehouses - 49312 | 504 | 3 |
| Natural Gas Liquid Extraction - 211112 | 450 | 3 |
| Other Chemical and Allied Products - 42269 | 356 | 2 |
| Farm Product Warehousing - 49313 | 326 | 2 |
| Liquified Petroleum Gas Dealers - 454312 | 307 | 2 |
| Support Activities for Crop Production - 11511 | 283 | 2 |
| Plastics Material and Resin Production - 325211 | 250 | 2 |
| Basic Organic Chemical Manufacturing (Not otherwise specified) - 325199 | 244 | 2 |
| Poultry Processing - 311615 | 216 | 1 |

---

[13] EPA has prepared "Model RMPs," for certain industries such as chemical distributors, warehouses, ammonia refrigeration, and sewage (wastewater) treatment. Typical processes in these industries are very similar making standardized accident prevention programs possible. The OCA information for a process in one of these industries could be compared to the model. However, the model does not characterize a typical facility nor would this provide a comparison of distances to endpoint submitted by other sources in that industry.

Table 7 shows how the reported industry sectors are distributed across all RMP processes. For example, 315 industry sectors (or 67% of the total number of industry sectors reported in the RMP database) have 5 or fewer RMP reporting processes within that sector. Likewise, 452 industry sectors (or 92% of the total) have 50 or fewer RMP reporting processes within that sector. As above, using the data in **Table 7**, if the public had access to the OCA information in 10 RMP processes, then comparisons could be made across 76% of all industry sectors. However, this would not be enough access to allow full industry comparisons for 113 industry sectors or 14% of the industry sectors.

**Table 7 – Distribution of RMPs Across NAICS Codes**

| Number of RMP Processes | Number of Industry Sectors (NAICS codes) having this number of processes. | Percentage of all NAICS codes having this number of processes |
|---|---|---|
| 1 - 5 | 315 | 67% |
| 1 - 10 | 362 | 76% |
| 1 - 15 | 389 | 82% |
| 1 - 20 | 403 | 85% |
| 1 - 50 | 435 | 92% |
| 1- 100 | 452 | 95% |
| 1- 4,034 | 475 | 100% |

### C.    Comparisons by Specific Chemical

Another way to compare facilities and processes is to compare the OCA results for the same chemical across different facilities. For example, someone may want to compare the OCA data for a local facility that uses ammonia to all other facilities that use ammonia. Or, someone may want to examine the OCA data for facilities using the same chemical that had an accidental release in their five-year accident history. **Table 8** lists the chemicals most often reported in worst-case scenarios and in the RMP five-year accident history. According to this Table, ammonia OCA data from 7,506 other facilities would need to be accessed. Similarly, if someone wanted to compare the OCA data for facilities that had accidents involving ammonia, then ammonia OCA data from more than 600 other facilities would need to be accessed.

**Table 8 – Chemicals Most Often Reported in RMP WCS and 5-Year Accident History**

| Chemical | Number of WCS Reported | Percent of all WCSs | Number of Accidents | Percent of accident history |
|---|---|---|---|---|
| Ammonia | 7506 | 45 | 656 | 34 |
| Chlorine | 4104 | 24 | 518 | 27 |
| Propane | 1228 | 7 | 54 | 3 |
| Flammable Mixtures | 847 | 5 | 99 | 5 |
| Sulfur Dioxide | 445 | 3 | 48 | 3 |
| Ammonia (aqueous) | 275 | 2 | 43 | 2 |
| Hydrogen Fluoride | 211 | 1.2 | 101 | 5 |

Another useful comparison is to compare the RMP and OCA data for a facility of interest with RMPs and OCA data for facilities in industry sectors that had an accident history. Accident history can be a useful indicator of the likelihood of release. **Table 9** presents the 21 industry sectors that reported the greatest number of accidents in the five-year accident history section of the RMP. This data, along with the number of facilities reporting in **Table 6**, provides a measure of likelihood that a facility in a particular sector will have an accident. For example, there is approximately one reported accidental release in the five year RMP accident history for every 50 facilities reporting in the Agricultural Supply sector and less than one for every 10 facilities reporting in the Sewage Treatment sector, while in the Petroleum Refineries sector there is more than one per reporting facility.

**Table 9 – RMP 5-Year Accident History by Industry Sector**

| Industry Sector Description - NAICS Code | Number of Accidents | Percent of RMP Accident History |
|---|---|---|
| Petroleum Refineries - 32411 | 190 | 10 |
| Water Supply and Irrigation - 22131 | 116 | 6 |
| Sewage Treatment - 22132 | 110 | 6 |
| Basic Inorganic Chemical Manufacturing (Not otherwise specified) - 325188 | 89 | 5 |
| Basic Organic Chemical Manufacturing (Not otherwise specified) - 325199 | 89 | 5 |
| Other Chemical and Allied Products - 42269 | 87 | 5 |
| Agricultural supply - 42291 | 85 | 4 |
| Alkalies and Chlorine Manufacturing - 325181 | 80 | 4 |

| | | |
|---|---|---|
| Nitrogenous Fertilizer Manufacturing - 325311 | 68 | 4 |
| Poultry Processing - 311615 | 67 | 4 |
| Petrochemical Manufacturing - 32511 | 55 | 3 |
| Pulp Mills - 32211 | 54 | 3 |
| Refrigerated Warehousing - 49312 | 50 | 3 |
| Animal (except Poultry) slaughtering - 311611 | 47 | 2 |
| Natural Gas Liquid Extraction - 211112 | 34 | 2 |
| Plastics Material and Resin Manufacturing - 325211 | 34 | 2 |
| Frozen Fruit, Juice, and Vegetable Manufacturing - 311411 | 32 | 2 |
| Meat Processed from Carcasses - 311612 | 31 | 2 |
| Paper (except newspaper) mills - 322121 | 25 | 1 |
| Industrial Gas Manufacturing - 32512 | 24 | 1 |
| Other Basic Organic Chemical Manufacturing - 32519 | 24 | 1 |

## CONCLUSION

The collection of enough information to adequately compare the hazards and risks between facilities is a critical element in understanding the risk and eventual risk reduction at a particular facility. Even though OCA data does not provide a complete picture of risk, a comparison of the worst-case scenario and alternative scenario (in particular the distance to endpoint for these scenarios) allows a greater understanding of the maximum extent (and a more likely extent) of the hazard component of the risk associated with RMP reporting facilities. Several types of comparisons are outlined in this chapter, and each one requires access to a significant number of RMPs and OCA data in order to generate a complete comparison.

# CHAPTER 7

# WHAT IS THE PUBLIC'S ACCESS TO OCA INFORMATION UNDER CSISSFRRA?

This assessment has so far examined whether public access to TRI and other risk information has resulted in risk reduction. It has also explored whether the nature of, and the means of access to, the information has affected the public's use of the information. With respect to OCA data in particular, the assessment has considered whether public access to that data would likely result in chemical accident risk reduction.

EPA's charge is to assess the incentives created by public access to "off-site consequence [OCA] information," which is related to, but not synonymous with, OCA data. CSISSFRRA defines "[OCA] information" as "those portions of a [RMP], excluding the executive summary of the plan, consisting of an evaluation of 1 or more worst-case release scenarios or alternative release scenarios, and any electronic data base created by the Administrator from those portions" (CAA section 112(r)(7)(H)(i)(III)). Notably, the definition refers to specified portions of RMPs and any EPA database created from these portions, not to the data reported in those portions. It also expressly excludes RMP executive summaries, which are required to include at least a brief description of the information reported in the OCA portions of RMPs. In fact, most RMPs include at least some OCA data in their executive summaries.

Relatedly, CSISSFRRA states that it "does not restrict the dissemination of [OCA] information . . . in any manner or form except in the form of a [RMP] or electronic data base created by [EPA]" (emphasis added) (CAA section 112(r)(7)(H)(xii)(II)). Together these provisions make clear that "OCA information" refers only to the specified forms of OCA data, not the data itself. (The statute restricts these forms of OCA data because they are formatted in a way that makes them fairly easy to compile into a large OCA database that could be posted on the Internet.) Accordingly, EPA must ultimately assess the incentives that would be created by public access to "OCA information," or OCA data in the restricted forms (sections 2 through 5 of RMPs and any electronic database EPA creates from those sections).

CSISSFRRA provides the public with some means of access to OCA data (i.e., the data reported in the OCA sections of RMPs) even before, or in addition to, the regulations that are to govern distribution of "OCA information." As noted above, RMP executive summaries are excluded from the definition of "OCA information" and thus from the statute's restrictions on dissemination. Executive summaries are required to include a "brief description" of OCA data. EPA's rule does not define a "brief description," leaving facilities to make reasonable decisions as to what information to include. A random sampling of RMPs indicates that the amount of OCA data included in executive summaries varies from facility to facility, with some facilities providing nearly complete data and others providing little. However, most executive summaries provide at least some OCA data. The summaries are already available on the Internet through multiple web sites.

Besides executive summaries, CSISSFRRA ensured early public access to at least summaries of OCA data. It required virtually all covered facilities by February 1, 2000 to conduct a public meeting or post a public notice summarizing the OCA sections of its RMP (CSISSFRRA section 4). To date, the Federal Bureau of Investigation has received notification from approximately 5000 facilities that they have complied with this requirement. In addition, CSISSFRRA allows any facility to actually release the OCA sections of its RMP to the public without restriction, and once a facility has done so, the information is no longer restricted (CAA section 112(r)(7)(H)(v)(III)). To date, EPA has received notification from over 900 facilities that they have released their OCA information without restriction.

CSISSFRRA also allows governmental officials to communicate OCA data to the public, so long as they do so in a form that does not replicate the OCA sections of RMPs or EPA's OCA database. CSISSFRRA guarantees governmental officials access to "OCA information" for their "official use" (see CAA section 112(r)(7)(H)(iv) and (ii)(cc)-(ee)). Governmental officials, referred to as "covered persons" by the statute, include officers and employees of federal, state or local government or their agents or contractors, and officers and employees of state and local emergency response officials or their agents or contractors (see CAA section 112(r)(7)(H)(i)(I)). Emergency response officials include members of State Emergency Response Commissions (SERCs) and Local Emergency Planning Committees (LEPCs) created under the Emergency Planning and Community Right-to-Know Act (EPCRA). Members of these commissions and committees can include members of the public, the media, and industry, as well as representatives of emergency responders such as fire and police departments (EPCRA section 301(c)). Emergency response officials, including fire fighters, are "covered persons" whether or not they are paid for their services.[14]

While CSISSFRRA guarantees covered persons access to OCA information, it prohibits them from disclosing the information to the public except as authorized by the statute or the regulations issued under it (section 112(r)(7)(H)(v)). It also prohibits them from disclosing "any statewide or national ranking of identified stationary sources derived from" OCA information. Any covered person who violates the prohibition is subject to criminal penalties of up to $1,000,000 for violations committed in any one year.

Notwithstanding these prohibitions, CSISSFRRA "does not restrict the dissemination of [OCA] information by any covered person in any manner or form except in the form of a [RMP] or an electronic data base created by [EPA] from off-site consequence analysis information," as noted above. Thus, CSISSFRRA prohibits disclosure of RMP sections 2 through 5, or OCA data conveyed in the "form" of those sections, and prohibits disclosure of EPA's OCA database. But it does *not* prohibit disclosure of OCA data when the data is disclosed in a form different than that

---

[14] "Covered persons" also include "qualified researchers" under CAA section 112(r)(7)(H)(vii).

43

of RMP sections 2 though 5 or EPA's OCA database. Congress so limited the scope of the prohibition so that governmental officials could communicate to the public about the potential off-site consequences of chemical accidents, but in a way that does not lend itself to Internet dissemination. (See, e.g., 145 Cong. Rec. H6083 (July 21, 1999) (statement of Rep. Dingell).) Covered persons may consequently convey to the public the data in RMP sections 2 through 5 and EPA's OCA database, so long as they do not hand out copies of, or otherwise replicate, the restricted RMP sections or provide access to EPA's database.

Finally, CSISSFRRA guarantees public access to OCA information itself (i.e., the OCA sections of RMPs and EPA's database created from those sections) in several specified ways apart from the regulations governing distribution of OCA information. First, it requires EPA to make OCA information available to the public without information concerning the identity and location of facilities reporting the information (CAA section 112(r)(7)(H)(iv)). That information is currently available upon request. Second, it requires EPA, in consultation with DOJ and other agencies, to establish a "read-only information technology system" that "provides for the availability to the public of [OCA] information by means of a central data base under the control of the Federal Government that contains information that users may read, but that provides no means by which an electronic or mechanical copy of the information may be made" (CAA section 112(r)(7)(H)(viii)). EPA is working with other federal agencies to develop this read-only system. Third, CSISSFRRA requires EPA, in consultation with DOJ, to make OCA information available to "qualified researchers" by means of a system that does not allow researchers who receive the information to disseminate it (CAA section 112(r)(7)(H)(vii)). EPA expects to initiate this system soon.

Against this backdrop of guaranteed but limited access, EPA must evaluate "the incentives created by public disclosure of [OCA] information for reduction in the risk of accidental releases[.]" EPA has so far assessed the incentive that public access to OCA data would create for chemical accident risk reduction. EPA must now consider the importance of public access to OCA information – the restricted forms of OCA data – to the creation of that incentive, in light of the public access CSISSFRRA already provides.

Guaranteed access to OCA information has several important advantages over the access to OCA data that CSISSFRRA otherwise provides. To begin with, OCA information provides full OCA data for covered facilities. RMP executive summaries, by contrast, communicate only as much OCA data as facilities choose to include. While a minority of facilities included nearly complete OCA data, others provided little. A person interested in a particular facility may or may not find the OCA data of interest to that person in the facility's executive summary.

In the public meeting or notice required by CSISSFRRA, facilities were required to summarize OCA data, not provide OCA data itself. Further, facilities had discretion regarding how to summarize OCA data.

Under CSISSFRRA, governmental officials are another potential source of OCA data, but whether they communicate it is left to their discretion. EPA is so far unaware of any

44

governmental officials communicating OCA data to the public. Indeed, few governmental officials have even requested access to OCA information to date. EPA has learned from several members of LEPCs and other state and local officials that they are reluctant to obtain OCA information out of concern for the criminal penalties associated with unauthorized distribution of it. CSISSFRRA only punishes willful violation of its disclosure restrictions, but these officials have nonetheless expressed concern that inadvertent disclosure might result in criminal fines. (32) In any event, if governmental officials do not have OCA information themselves, they cannot communicate OCA data to the public. More fundamentally, providing governmental officials with discretion to convey OCA data leaves up to them whether and to what extent they provide access to the data. And even if government officials choose to communicate OCA data, the public cannot be sure they were communicating it accurately without access to OCA information itself.

Facilities may also disclose their OCA data to the public, but they, too, have discretion regarding whether to do so. Questions of accuracy may also arise to the extent facilities release something other than the OCA sections of their RMPs. To date, only a small percentage of facilities have released the OCA sections of their RMPs without restriction.

Given these limitations on existing public access to OCA data, guaranteed access to OCA information would assure members of the public that they can receive complete and accurate OCA data for the facilities of interest to them.

Guaranteed access to OCA information would also ensure that the public could gain access when they needed it. Governmental officials and facilities may or may not choose to communicate OCA data at any given time. CSISSFRRA required facilities to provide only one public meeting or notice by a date now past. Initial indications are that relatively few members of the public attended the meetings. For persons who missed the meeting or notice or moved to the area afterwards, there is no subsequent opportunity to obtain at least a summary of OCA data from the facility itself.

OCA information, by virtue of its format, has the further advantage of putting OCA data into context within an RMP. When presented as part of the RMP, OCA data can be reviewed together with information about a facility's accident history and prevention program. When considered along with RMP information, OCA data provides greater insight into the risk a facility poses and the steps the facility is taking to manage or reduce that risk.

In addition to providing some access to OCA data, CSISSFRRA guarantees that OCA information itself will be available to the public in several prescribed ways independent of the regulations. However, these avenues of access have significant drawbacks. OCA information without facility identification or location information is of limited utility. The most obvious limitation is that such a database does not allow a member of the public to know which facility's data he or she is viewing. A more subtle limitation has to do with the fact that RMP reporting is general in nature. RMPs do not explain how a facility is using prevention measures to reduce risk, only that certain types of measures are being used. For example, a facility need only report that it uses shut-off valves and a dike to reduce the volume and dispersion of an accidental release, but it does not have to provide the specifications or locations of those measures. For a member of the public to fully understand a facility's prevention program, he or she must contact the facility (or ask a governmental official or other representative to do so) to gain more specific information. The limited usefulness of OCA information without facility identification data is perhaps best indicated by the fact that, to date, no one has requested the information.

The read-only OCA database required by CSISSFRRA is also likely to have drawbacks. It is unclear at this time how such a read-only system will operate and how many outlets for the system will be available. Government resources for the read-only technology system are limited. Finally, the required system for access to OCA information for qualified researchers is, by definition, not available to the general public. While the system may result in risk information being developed and communicated to the public by researchers, it does not provide the public with access to OCA information itself and it precludes researchers from providing that access.

In summary, CSISSFRRA already makes OCA data, and even OCA information (the restricted forms of OCA data), available to some extent. EPA has thus evaluated the incremental benefit of guaranteeing public access to OCA information. While the public has access to at least some OCA data through RMP executive summaries and may obtain additional OCA data as a result of other, discretionary actions by governmental officials or facilities, the extent to which the public receives OCA data through these avenues ultimately depends on decisions made by others. CSISSFRRA's provisions for public access to OCA information without facility identification information and in read-only database are limited in nature and thus effect. Without guaranteed access to OCA information as such, a member of the public may or may not obtain the OCA data for the facilities of interest to her or him. And without that data, she or he would have no basis – or incentive – to initiate actions directed at reducing the risk of chemical accidents. Furthermore, a lack of data on the part of the public would mean that facilities, including those that lag behind their peers, would have fewer incentives to reduce the risk of chemical releases.

# CHAPTER 8

# FINDINGS

The President delegated to EPA the responsibility for assessing the incentives for reduction in chemical accidents created by public disclosure of OCA information. This assessment has produced a number of findings that address the relevant issues.

- Chemical accidents impose substantial costs on the American public and on industry. Although catastrophic chemical accidents that kill many people at once are fortunately relatively rare, chemicals accidents continue to cause deaths, injuries, property damage, disruption of lives, and business losses. These accidents continue to occur at facilities that are subject to accident prevention regulations. Almost 80% of the serious accidents reported in the RMP five-year accident history occurred at facilities subject to the OSHA PSM standard. Although it is likely that the PSM standard has prevented some releases, there is clearly a need for additional efforts to improve safety at facilities handling highly hazardous chemicals.

- Public information does result in the public acting on the information, and those actions have very likely led to risk reduction. Public use of the TRI data has included reports and campaigns from public interest groups at the local, state, and national level, press coverage, and state legislation. It is not possible to quantify the level of risk reduction produced by such actions with any certainty. TRI emissions have decreased by 43% since 1988, although other factors have produced some of that reduction. Negative press coverage directed at certain facilities appears to have led these facilities to achieve reductions in their TRI emissions.

  Over the last 10 years, reportable releases of hazardous substances from facilities in four states reviewed have declined by 68 percent while reports of oil spills and transportation releases have showed no consistent trend. All the reasons for the decline in reportable releases are not known, but one likely factor is that hazardous substances are the subject of public scrutiny to a far greater degree than oil and transportation, leading facilities to improve their management of these chemicals to reduce risks.

- The type of information provided to the public affects its use. Scorecard shows that the public is at least an order of magnitude more likely to access interpreted data than it is to seek raw data. This is because interpreted data, such as OCA data, are put into context and made more understandable by the public.

- Ease of access is important. The greater the effort needed to obtain data, the less likely members of the public will obtain it. EPCRA section 312 data that are available locally have rarely been used (about 3,500 requests per year). TRI and related data on the

Internet, even without interpretation, draw 240,000 searches a year at one Internet site. With interpretation and easy access, Internet users view TRI and other data over seven million times per year at another website.

- Current public access to OCA data is uneven, and consequently is ineffective relative to consistent, broad access. Executive summaries vary widely in the amount of OCA data reported. Public officials have been reluctant to even access the data because of liability concerns; there has thus been little chance for the public to learn OCA data from them.

- OCA could be derived, though not accurately, from the raw data, but the technical challenges are likely to prevent the public from doing so.

- Actual chemical releases are different from the releases evaluated in the OCA information, but, because facilities generate OCA information according to consistent assumptions, the public and others can understand and compare the relative potential hazards and risks present at a facility.

As noted at the outset of this assessment, chemical accidents continue to claim lives, health and property. The evidence and analysis set forth in the assessment demonstrate that providing the public with reasonably convenient access to OCA information would likely result in significant reductions in the risk of chemical accidents over and above the risk reduction that the RMP rule on its own will accomplish. Conversely, to the extent the public is not given workable access to OCA information, the risk of chemical accidents is not likely to be reduced as much as it world if such access were provided. And the loss of that increment in risk reduction would mean the loss of lives, health and property that could have been saved.

# REFERENCES

1.   *Loss Prevention in the Process Industries*, Second Edition, F.P. Lees, Butterworth-Heinemann Publishing, 1996, App. 5/9.

2.   *Loss Prevention in the Process Industries*, App. 4/2.

3.   Clean Air Act, 42 U.S.C. section 7412 (r).

4.   U.S. Chemical Safety and Hazard Investigation Board, "Chemical Safety Board To Conduct Comprehensive Investigation of Pennsylvania Chemical Plant Explosion," March 3, 1999.

5.   Economic Analysis in Support of Final Rule on Risk Management Program Regulations for Chemical Accident Release Prevention, As Required by Section 112(r) of the Clean Air Act, May 21, 1996.

6.   Adams, W., S. Burns, and P. Hardwerk.  National LEPC Survey, October 1994.  George Washington University, App. A, Table 25.

7.   U.S. Government Accounting Office, "Environmental Information: Agencywide Policies and Procedures Are Needed for EPA's Information Dissemination," GAO/RCED-98-245, Sept. 1998.

8.   Lynn, F. M. and J. Kartez. "Environmental Democracy in Action: the toxics release inventory," *Environmental Management*, Vol. 18, No. 4, pp. 511-521, 1994.

9.   *1997 Toxics Release Inventory*, US EPA, EPA 745-R-99-003

10.  World Bank, "Greening Industry: New Roles for Communities, Markets, and Governments," Oxford University Press, 1999, pp 64-74.

11.  "With Toxic Risk, Plans Vary," Washington Post, Oct. 10, 1999, p C1, C10-C11.

12.  "Plant Warnings Go Unheeded," Washington Post, Nov. 5, 1999, p A1.

13.  "Blue Plains Details Safety Plans," Washington Post, March 3, 2000, p B3.

14.  Robert Barrish, Delaware Department of Natural Resources and Environmental Control, Division of Air & Waste Management, personal communication, with David Wiley, EPA, Dec. 20, 1999.

15.  Mark Zusy, Nevada Department of Environmental Protection, Bureau of Waste

Management; personal communication with Craig Matthiessen, EPA, January 2000.

16.    National Institute of Chemical Studies, *NICS News,* Vol. 8, Fall 1999, p 6.

17.    National Highway Traffic Safety Administration, "Consumers need more reliable automobile fuel economy data," Report No. CED-81-133, July 29, 1981

18.    Henderson, David., "The Economics of Fuel Economy Standards," RGO, Vol. 9, No. 1, Jan./Feb. 1985, pp 45-48.

19.    Neuhouser,  Kristal, and Patterson. "Use of food nutrition labels is associated with lower fat intake," Journal of the American Dietetic Association, 99, 1, 45(6), Jan., 1999.

20.    Correspondence with Dr. William S. Pease, Director, Internet Projects, Environmental Defense Fund, 11/22/99 and 2/22/00.

21.    Wolf, Sidney.  "Fear and loathing about the public right to know: the surprising success of the Emergency Planning and Community Right-to-Know Act." *Journal of Land Use and Environmental Law*, V. 11, No. 2, Spring 1996.

22.    Adams, W., S. Burns, and P. Hardwerk.

23.    Reginald Baldini, NJ Department of Environmental Protection, personal communication, with David Wiley, EPA, Dec. 8, 1999.

24.    Robert Barrish, Dec. 20, 1999.

25.    Mark Zusy, January 2000.

26.    *Technical Guidance for Hazards Analysis - Emergency Planning for Extremely Hazardous Substances*; US EPA, US FEMA, US DOT; December 1987.

27.    *Planning for the Worst, Maps*; The Augusta Chronicle Online, 10/9/97; Augusta, Georgia; www.augustachronicle.com.

28.    *Loss Prevention in the Process Industries*, Second Edition, F.P. Lees, Butterworth-Heinemann Publishing, 1996, pp 2/2-3.

29.    *Safety in the Chemical Industry, Lessons from Major Disasters*, E.A. Stallworthy and O.P. Kharbanda, G.P. Publishing, Columbia, MD, 1988, pp 99-100.

30.    Guidelines for Evaluating the Characteristics of Vapor Cloud Explosions, Flash Fires, and BLEVEs, CCPS/AIChE, 1994, pp 70-75.

31.    *Safety in the Chemical Industry*, pp 99-100.

32.    Timothy Gablehouse, Chair, Jefferson County Colorado LEPC and LEPC representative
       to the Federal Accident Prevention Advisory Subcommittee, personal communication with
       David Wiley, EPA, Dec. 2, 1999.

33.    Graham, Mary; *Regulation by Shaming*; The Atlantic Monthly, April 2000.

34.    *Right-to-Know Planning Guide*, "EPA Notices Change in Users of Toxic Release
       Inventory." Bureau of National Affairs. (February 1990): p. 4.

35.    Tryens, Jeffrey, Richard Schrader, and Paul Orum. *Making the Difference: Using the
       Right-to-Know in the Fight Against Toxics*. Washington, DC: Center For Policy
       Alternatives and Working Group on Community Right-to-Know. (undated): p. 2.

36.    ICF, Inc., Memorandum: "Transmittal of Research Materials on Negative Press," April
       18, 2000.

This page intentionally left blank.

**APPENDIX A**

**BACKGROUND
OF
EPA'S RISK MANAGEMENT PROGRAM
AND
OCA INFORMATION**

# APPENDIX A

# BACKGROUND OF EPA'S RISK MANAGEMENT PROGRAM AND OCA INFORMATION

This Appendix provides background information on the accident prevention provisions of the Clean Air Act, describes EPA's Risk Management Program and the information contained in a Risk Management Plan (RMP), and discusses what Off-site Consequence Analyses are, including what information is included in the OCA portion of a risk management plan, and what is not included.

## 1. What Role Do the Clean Air Act Amendments of 1990 Play in Accident Prevention?

The CAA Amendments of 1990 authorized regulations and programs to prevent accidental chemical releases and to minimize the consequences of such accidental releases when they occur. The sections added to the CAA for this purpose are:

- Section 112(r)(1) - establishes a general duty on facilities handling any extremely hazardous substance to identify hazards which may result from accidental releases using appropriate hazard assessment techniques, to design and maintain a safe facility, and to minimize the consequences of accidental releases which do occur.

- Sections 112(r)(3), (4) and (5) - require EPA to establish a list of at least 100 substances that pose the greatest risk of causing death, injury, or serious adverse effects to human health or the environment from accidental releases along with a threshold amount for each substance.

- Section 112(r)(6) - establishes a Chemical Safety and Hazard Investigation Board to investigate accidental releases and advise EPA and the Department of Labor, Occupational Safety and Health Administration (OSHA) on the efficacy of their regulatory programs.

- Section 112(r)(7) - directs EPA to issue reasonable regulations and appropriate guidance to prevent and detect accidental releases and to require facilities[1] with more than a threshold amount of a substance listed under sections 112(r)(3)-(5) to develop and implement risk management plans (RMPs). This section specifies that RMPs include an evaluation of the off-site consequences of worst-case releases; the RMPs, including the consequence evaluations, must be submitted to the Chemical Safety Board and state and local officials and be made available to the public.

---

[1] The CAA and the regulations use the term "stationary source" rather than "facility," which is used in EPCRA. These terms are synonymous; this report generally uses the term "facility."

- CAA Amendment section 304 - requires the Department of Labor to promulgate regulations under the Occupational Safety and Health Act to establish a chemical process safety management (PSM) standard designed to protect employees from hazards associated with accidental releases of highly hazardous chemicals in the workplace. This standard was issued by OSHA in 1992. It requires facilities having more than a threshold quantity of certain highly hazardous chemicals listed by OSHA to implement and document an accident prevention program. For example, facilities subject to OSHA PSM must prepare and use written operating and maintenance procedures for hazardous chemical processes, must conduct a systematic analysis of process hazards and ensure that all hazards are controlled, and must conduct training for workers who operate hazardous chemical processes. The elements of the OSHA PSM accident prevention program also serve as the core accident prevention elements for EPA's Risk Management Program.

## 2.    What are the Requirements of EPA's Risk Management Program Regulation?

EPA addressed the chemical accident prevention and detection requirements in 112(r)(7)(B)(i) along with the risk management plan requirements in (B)(ii) in one regulatory effort. In so doing, EPA required that facilities handling more than the threshold quantity of a substance listed under sections 112(r)(3)-(5) must develop and implement a Risk Management Program that evaluates the hazards present at the facility and establishes a chemical accident prevention and emergency response programs. The most important feature of the Risk Management Program is the chemical accident prevention program which is based on the elements of chemical process safety management. The most effective way to prevent catastrophic chemical accidents is through process safety management. The elements of process safety management were derived by industry, trade associations, and professional societies. Chemical process safety management calls for a systematic and rigorous evaluation of the chemical and process hazards present in the facility and it brings together all the necessary elements for the safe operation of that process, day-after-day, under one management system. Through process safety management, facilities design, install, maintain, and operate the equipment necessary to prevent and detect accidental releases.

The elements of process safety management were adopted by OSHA into its PSM standard promulgated in 1992. EPA chose to adopt and build on OSHA's requirements and the industry approach for its chemical accident prevention program because it is the most effective way to prevent accidents, most chemical accident prevention actions taken to protect workers will also protect the public, and separate requirements would be more burdensome, duplicative and less proven than process safety management.

While EPA's Risk Management Program borrows much from the OSHA PSM standard, its requirements extend beyond the OSHA standard. In addition to the PSM accident prevention requirements, the EPA Risk Management Program requires facilities to: analyze the off-site consequences of accidental releases, provide information about accidents that the facility has suffered during the previous five years, and provide information about their accident prevention and emergency response programs in a publicly-available risk management plan (RMP). These additional elements were intended, in large part, to use "the power of public scrutiny to promote voluntary hazard reduction, often achieving far more benefits than what regulatory programs could achieve on their own" (Legislative history of the Chemical Safety Information, Site Security, and Fuels Regulatory Relief Act, Senate Report 106-70, June 9, 1999, page 12). By requiring companies to analyze the potential off-site consequences of accidental releases and including this analysis in the RMP, Congress and EPA predicted that the "right-to-know effect," when applied to the chemical industry by means of publicly available RMPs, would contribute to an atmosphere in which industry, through non-regulatory means, sees incentives to take all reasonable steps to reduce chemical risks.

In the Preamble to the final RMP rule published in the *Federal Register* on June 20, 1996, EPA discussed the two underpinnings of EPA's approach to the risk management program. First, EPA stated that with this rule "EPA continues the philosophy that the Agency embraced in implementing the Emergency Planning and Community Right-to-Know Act of 1986. Specifically, EPA recognizes that regulatory requirements by themselves, will not guarantee safety. Instead, EPA believes that information about hazards in a community can and should lead public officials and the general public to work with industry to prevent accidents. EPA intends that officials and the public use this information to understand the chemical hazards in the community and then engage in a dialogue with industry to reduce risk."

Secondly, the Agency stated that the rule "builds upon existing programs and standards. For example, EPA coordinated with OSHA and the Department of Transportation (DOT) in developing this regulation. To the extent possible, covered sources will not face inconsistent requirements under these agencies' rules." This approach was supported by public comments which stated that sound process safety management systems ideally address chemical accident prevention in a way that protects workers, the public and the environment.

In the *Federal Register* on January 31, 1994, and in subsequent amendments, EPA published a list of regulated substances under the risk management program. This list currently contains 77 substances listed because of their volatility and acute toxicity, and 63 substances listed because of their high flammability. Together, these substances represent the initial list of regulated substances that are known to cause, or may be reasonably anticipated to cause, death, injury, or serious adverse effects to human health or the environment if accidently released.

In the same notice, EPA published threshold quantities for each of the regulated substances. The approach taken to set the threshold quantities focuses on the quantity of a substance that might be released in a single accident, and that could be reasonably anticipated to cause severe health effects as a result of an accidental release. Threshold quantities for toxic substances range between 500 to 20,000 pounds. The threshold quantity for all flammable substances is 10,000 pounds.

Any facility having more than a threshold quantity of a regulated substance in a process is required to develop and implement a risk management program. However, EPA scaled the risk management program requirements according to the relative risk posed by a facility to the surrounding community. Each process at a facility that has had no serious accidents in the past five years and that can demonstrate that the worst-case accident scenario does not affect any public receptors is eligible for "Program 1" requirements, which are the least stringent. Processes at facilities not eligible for Program 1 that are already subject to OSHA PSM or belong to industry sectors having a significant history of accidents are required to meet "Program 3" requirements, which are the most extensive. Finally, processes that are not eligible for Program 1 or subject to Program 3 are required to meet "Program 2" requirements. Program 2 requirements are similar to Program 3 except that they contain streamlined accident prevention program measures (the OCA, accident history, and emergency response program requirements are the same in Programs 2 and 3).

The risk management program must include:

- A management system - Owners or operators of facilities with Program 2 or Program 3 processes must develop a system to oversee the implementation of the risk management program elements, and assign a qualified person or position that has the overall responsibility for development, implementation, and integration of the program elements. Program 1 processes are not subject to the management system requirements.

- An accident prevention program - For Program 3 processes, this is virtually identical to the requirements under OSHA PSM, including measures such as written operating and maintenance procedures, process hazard analysis, a mechanical integrity program, incident investigations, compliance audits, and others. For Program 2 processes, the accident prevention program contains a streamlined subset of the full PSM requirements. Program 1 processes are not subject to any additional prevention program requirements. Prevention program requirements for Program 2 and 3 processes are "performance-based," as opposed to "command-and-control," because chemical facilities and processes are unique and therefore accident prevention programs must necessarily be tailored to each facility.

- An emergency response program - requires facilities to have procedures in place to notify emergency response officials in the event of an accident and to coordinate with local response agencies and community response plans. If a facility responds to its own emergencies, facilities must have a written plan containing procedures for informing emergency response agencies about emergencies, documentation of proper first-aid and emergency medical treatment, procedures for emergency response to an accidental release, including use of equipment, and training for employees who will respond to releases.

- A hazard assessment program that consists of a five-year accident history and an analysis of the consequences of worst-case and other accidental releases:

  1. The five-year accident history includes a description of prior accidental releases which meet certain severity triggers, such as deaths, injuries, significant property damage, evacuations, environmental damage, or sheltering in place.

2. The off-site consequence analysis (OCA) includes an analysis of the potential consequences of hypothetical worst-case and alternative release scenarios. Worst-case scenarios assume the release of the greatest amount of a regulated substance held in a single vessel or pipe under specified ambient and process conditions, taking into account administrative controls that limit the maximum quantity, and accounting for the effects of passive mitigation features if present. Alternative release scenarios assume a release that is more likely to occur than the worst-case, using release parameters chosen by the facility owner as appropriate for the scenario, and may account for both passive and active mitigation features.

As required by Congress in section 112(r), all facilities subject to the RMP regulation, regardless of Program level, must submit an RMP that documents elements of their risk management program. RMP contents are described in detail in the next subsection.

Facilities subject to the program were required to submit an initial RMP by June 21, 1999, and must submit an update at least every five years. Facilities must update their RMP sooner under certain circumstances (e.g., major change involving a regulated substance or process at the facility). Approximately 15,000 facilities have submitted an RMP to date.

## 3. What information is reported in a Risk Management Plan?

The risk management plan (RMP) is intended to provide information that can be used by others to judge the risk that a facility poses to the surrounding community and to understand the steps taken by that facility to manage that risk. (**A fictitious sample RMP is shown in section 15 below**.) The executive summary is an overall text description of a facility's risk management program, including, in general terms, the potential off-site consequences of the accidental releases from the facility. The rest of the data in the RMP generally consists of yes/no, check-off box, and numerical answers to standard questions. There are additional areas where facilities may include text explanations for various entries, but (with the exception of the executive summary) these are optional. The advantage of this format is that it allows data to be easily submitted, compiled and managed in electronic form. However, also as a result of this format, information submitted in RMPs is usually not extremely detailed. For example, a facility would indicate, by checking various choices in a list, what types of mitigation measures it uses in a process, but unless the facility chooses to add an optional explanation, the reader can not discern details such as precise locations, methods of operation, or design features of those devices. Facilities are, however, required to maintain on-site documentation which supports the information contained in the RMP and implementation of the overall risk management program.

RMPs contain the following sections. The presence of some sections and the total number of pages for RMPs vary depending on the number and type of processes and chemicals present at a facility.

- Section 1: Registration information (e.g., facility name, address, process chemicals, etc.) and an executive summary which provides a brief description of the accidental release prevention and emergency response policies at the source, worst-case and alternative case scenarios, five-year accident history, and planned changes to improve safety.

- <u>Sections 2-5</u>: Evaluation methodology and data for the off-site consequence analyses of worst and alternative release scenarios. These sections provide data on the possible consequences of the scenarios, as well as the assumptions and models used to obtain this data.

- <u>Section 6</u>: Five-year accident history data. For each past accident, the facility provides the date of the event, chemical(s) released, source of release, on-site and off-site impacts, initiating event, and factors contributing to the release.

- <u>Section 7</u>: Contains a description and data for the processes subject to prevention Program 3. Besides an optional narrative on the prevention program, facilities are required to provide such information as the date of the last process hazards analysis, the major hazards identified by that analysis, process controls used to address these hazards, and information on maintenance, training, compliance audits, and incident investigations.

- <u>Section 8</u>: Like Section 7, this section contains a description and data for processes subject to prevention Program 2 .

- <u>Section 9</u>: Contains data on the facility's emergency response program and plan.

## 4.     How is RMP Information Managed?  How Did EPA Intend for the Public to Obtain Access?

EPA learned from its experience with the EPCRA program (see Chapters 3 and 4) that electronic submission of data has several benefits over the submission of paper forms. First, electronic submissions reduce the burden on regulated and receiving entities. Second, the Agency noted that local agencies often lack the resources needed to make use of the information reported to them by industry under federal programs, so the Agency wanted to limit the information management burden on local entities so they could focus on the chemical safety issues raised by the data. Third, EPA learned that electronic submissions would benefit affected communities and the general public. The Agency believed this type of submission would promote consistency and uniformity to enable communities and the general public to better understand the data. When the agency proposed this method of data management, most of the public comments supported the proposal to submit RMPs in electronic form to a central location, and EPA adopted this approach when it promulgated the RMP rule (61 Fed. Reg. at 31673, 31694-95).

To help implement EPA's decision to develop a centralized management system for RMPs, EPA convened a subcommittee under the Federal Advisory Committee Act, which included representatives from state and local governments (including LEPCs), academia, industry and public interest groups. One of the major issues the Accident Prevention subcommittee considered was how RMPs should be submitted and how RMP information could be managed. The subcommittee unanimously agreed with EPA that RMPs should be submitted electronically, that EPA should compile the RMPs into a central electronic database, and that EPA should make that database available to Chemical Safety Board and state and local officials.

In a Supplemental Notice of Proposed Rulemaking, EPA requested comments on how public participation in the risk management program process could be encouraged. EPA's preferred approach was to encourage the public and sources to use existing groups, primarily LEPCs, as a conduit for communication between the source and the public throughout the RMP development process. While a substantial number of commenters supported this approach, many opposed it because some LEPCs are not functional and because LEPCs were not seen as an appropriate substitute for public participation. In the final rule, EPA did not adopt any specific public participation requirements, but rather decided at the time to make all of the RMP information immediately available to the public. EPA believed that by doing this, people would be able to compare facilities in their community with similar facilities in other areas, and thereby gain a better understanding of local industries in order to carry on a more informed dialogue with facilities about their hazards and accident prevention practices.

The Accident Prevention subcommittee also unanimously agreed that the electronic RMP database, except for the OCA information, should be made available to the public over the Internet. Most subcommittee members believed that the OCA portion of the database should also be available to the public on the Internet, but one member expressed concern that placing the OCA information on the Internet could provide a targeting tool for criminals and terrorists. Law enforcement agencies shared this concern, which evolved into Public Law 106-40. The public currently has unrestricted access via the Internet to all sections of the RMP database except the OCA information. EPA's database of OCA information is currently withheld from the public in accordance with Public Law 106-40.

Finally, as stated previously, more than any other information reported in an RMP, OCA information provides an easily understood means of evaluating the hazards a particular facility poses to its surrounding community and how its hazards compare to those of similar facilities. The data tell the public how far a worst-case or alternative scenario release from a particular facility could travel, roughly how many people could potentially be affected, and what types of "public receptors" (e.g., homes, hospitals, schools, businesses, parks) could be in the path of a release. In short, the data allow members of the public to determine whether they could be harmed by a chemical release from a particular facility.

## 5.    What is OCA information?

Off-site Consequence Analysis (OCA) information is the portion of an RMP that contains analyses of the possible consequences to a surrounding community of hypothetical chemical accident scenarios. While these scenarios are not meant to predict exactly what will occur in an actual event, they are an attempt to determine what could occur under certain conditions, and to present this information in an easily understood fashion.

The OCA is a *consequence* assessment, not a risk assessment. Risk assessments typically measure risk as the product of the hazard multiplied by its probability (or frequency). For example, the risk of being struck and killed by lightning is a function of the hazard (lightning's potential to cause death) and the likelihood that if you are outdoors during a thunderstorm, that you will be struck. Accidental release risk considers the hazard of the chemical accidentally released (e.g., it's toxicity), the consequences of that release (how much will you be exposed to it) and the likelihood that the chemical will be accidentally released. The OCA provides a rough estimate of only the hazards and consequences of an accidental release, without evaluating the likelihood or probability that such an accident will occur. Since the likelihood of an accidental release varies considerably from facility to facility, the RMP requirements, in effect, assume that the likelihood of a worst-case release or alternative case release is equal across all facilities. This simplifies the analysis and allows comparability of hazards and consequences across RMP facilities. The consequences are expressed in terms of the potentially affected population, as well as the types of buildings, parks, and other public and environmental areas that could be seriously affected by a release.

The OCA requirements of EPA's Risk Management Program uniquely distinguish it from any other federal regulatory program. Each facility subject to the RMP regulations must conduct an OCA, report the results of the analysis in their RMP submission, and in most cases, discuss the analysis with the local public. In an OCA, all sources are required to develop at least one *worst case scenario*. Additionally, most sources (all except Program 1 sources) must also develop at least one *alternative release scenario*.

## 6.    What is a worst-case scenario?

Worst case scenarios are highly unlikely accident scenarios that are intended to serve as a measure of the maximum hazard that a chemical facility could pose to the surrounding area. They assume that the facility accidentally releases the entire contents of its largest tank or pipe of toxic or flammable material into the environment under very stable atmospheric conditions. The scenario also assumes that any active release mitigation systems (i.e., systems that require human, mechanical, or energy input to function) fail to operate, but that passive systems do operate. Stable atmospheric conditions (i.e., low wind speed and high atmospheric stability) are assumed because they are most conducive to lengthening the distance that a highly concentrated toxic gas cloud travels as it moves outward from its source; breezy, unstable conditions cause the cloud to disperse in a relatively short distance. For flammable worst case scenarios, the analysis assumes that a vapor cloud explosion occurs after the release. A vapor cloud explosion is the type of flammable gas accident that could generally affect the greatest geographic area.

## 7.    What is an alternative release scenario?

Alternative release scenarios are more likely, and generally less severe, than worst case scenarios. These scenarios are intended to be a more realistic estimate of the consequences to the surrounding community of a chemical accident at a given facility. Many facilities and local emergency planners use them to prepare emergency response plans. Alternative scenarios are selected by the facility based on the facility's accident history, hazards analysis, or the experience and judgement of the owners or operators of the facility. Alternative scenarios incorporate more realistic assumptions in their analysis than worst-case scenarios. Alternative release scenarios would generally assume that a smaller accidental release occurs under typical atmospheric conditions and that passive and active release mitigation systems operate properly. For example, an alternative scenario might be a 10-minute leak from a split in a transfer hose on a breezy, partly cloudy day.

## 8. How are OCA scenarios developed and how accurate are they?

Both worst-case and alternative release scenarios are hypothetical estimates. They are analyzed using mathematical vapor cloud dispersion models (for toxic scenarios) or fire/explosion models (for flammable scenarios) to predict either the extent that a toxic gas cloud would spread or the extent of blast or radiant heat effects from an explosion or fire of highly flammable material. Each scenario includes an estimated distance outward from the source that may be subject to concentrations, over pressures, or high temperatures of a toxic substance release or flammable chemical explosion or fire that could cause irreversible acute health effects or death to human populations within that range. The analyses are based on estimates of the quantity of a chemical released, the rate of release, airborne dispersion and the airborne concentrations (for toxics) or blast effects (for flammable substances) that could cause at least irreversible health effects.

Many valid methods are available to conduct the OCA. These methods usually involve using either computer programs or lookup tables in which the analyst enters various parameter values (e.g., atmospheric conditions, terrain roughness, etc.), and the computer program or lookup table (also based on a computer simulation) provides the scenario endpoint, or consequence distance. For worst-case scenarios, most input assumptions are specified by regulation in order to provide some basis for comparison among similar sources. However, other assumptions are facility-dependent and are therefore selected by the facility analyst (for this reasons and others explained below, the OCA results from two similar sources may differ greatly). In alternative scenarios, all parameters are selected by the facility analyst.

The results of OCA scenarios are rough estimates that are generally conservative (i.e., likely to over-predict actual consequences). The uncertainty in the estimates arises from several factors, including the fact that actual atmospheric conditions at the time of an accidental release will be unknown, process conditions may change from those selected for analysis, and because the science of modeling large gas releases over long distances is highly complex and still evolving. Results will usually over-predict actual consequences for two reasons. First, the rule-specified assumptions for worst-case scenarios represent conditions that are most conducive to causing severe off-site effects. However, these conditions (i.e., high atmospheric stability and low wind speed for toxics, 10% explosive yield factor for flammable scenarios), are extremely rare. Any change from these conditions during an actual release will generally reduce the consequence distance. The second reason that OCA scenarios usually over predict consequences is because of the uncertainty in our understanding of some of the physical mechanisms involved in atmospheric dispersion and blast propagation. This has generally led scientists to err on the side of caution when developing models.

## 9.  Exactly what data appears in the OCA information of an RMP?

Facilities are required to analyze and report in the RMP one worst-case release scenario for each Program 1 process, one worst-case scenario for a toxic substance, one worst-case scenario for a flammable substance, and additional worst-case scenarios if a worst-case release scenario can affect a different public receptor then the other worst-case scenarios listed above. Sources are required to identify and analyze at least one alternative release scenario for each regulated toxic substance held in a covered process(es) and at least one alternative release scenario to represent all flammable substances held in covered processes.

Only certain data elements for these scenarios are reported in sections 2 through 5 of the RMP.  Section 2 is for toxic substance worst-case scenario data, Section 3 is for toxic alternative release scenario data, Section 4 is for flammable substance worst-case scenario data, and Section 5 is for flammable alternative release scenario data.  (See **Table A-1** for a list of the data elements and see the **Sample RMP** below.)

| Table A-1 – Data Reported in OCA Sections of an RMP | |
| --- | --- |
| **RMP Sections** | **Data Elements:** |
| 2.1, 2.2, 3.1, 3.2, 4.1, 5.1 | Chemical name, percent concentration, and physical state |
| 2.3, 3.3, 4.2, 5.2 | Dispersion model used to conduct the analysis (e.g. lookup table, RMP*Comp software) |
| 2.4, 3.4, 4.3, 5.3 | Release scenario (e.g., gas leak, liquid spill and vaporization, pipe leak, etc.) |
| 4.5, 5.5 | Consequence endpoint assumed (e.g., explosion over pressure, radiant heat level) (flammable scenarios only; toxic endpoints are mandated by rule) |
| 2.5, 2.6, 2.7, 3.5, 3.6, 3.7, 4.4, 5.4 | Quantity released, release rate, and release duration |
| 2.8, 3.8 | Wind speed (for worst-case, must be 1.5 meters/sec unless facility has other data) |
| 2.9, 3.9 | Atmospheric stability class (for worst-case, must be most stable [F] unless facility has other data) |
| 2.10, 3.10 | Topography of area surrounding the process or facility (urban or rural) |
| 2.11, 3.11, 4.6, 5.6 | Distance in miles to either the toxic or flammable endpoint |
| 2.12, 3.12, 4.7, 5.7 | Estimated residential population within the endpoint distance |
| 2.13, 3.13, 4.8, 5.8 | Public receptors (e.g., schools, residences, recreation areas, etc.) within the endpoint distance |

| 2.14, 3.14, 4.9, 5.9 | Environmental receptors (e.g., national or state parks, etc.) within the endpoint distance |
|---|---|
| 2.15, 3.15, 4.10, 5.10 | Passive mitigation considered (i.e., equipment that functions without human, mechanical, or energy input that is designed to limit a release) |
| 3.16, 5.11 | Active mitigation considered (alternative scenarios only) |
| --- | Graphics file name (optional). Facilities may include a map or other graphic to illustrate a release scenario |

## 10. What data does *not* appear in OCA information?

It is important to note what data does *not* appear in OCA information (or any other portion of an RMP). **OCA information does *not* contain the following**:

- **The number of people that would be killed by a worst-case scenario.** While the OCA information contains estimates of affected populations inside worst case and alternative release scenario circles, these are not estimates of the number of fatalities or injuries that would occur following such a scenario. As described above, the population reported for a worst-case scenario includes the total population inside a circle whose radius is equal to the distance to a particular endpoint. The endpoint is the concentration of toxic substance in a cloud (for toxic substances) or the radiant heat or overpressure (flammable substances) beyond which someone could be exposed for a short time and suffer no serious irreversible injury. However, since toxic gas clouds generally travel in the direction of the prevailing wind, they form a long, narrow plume, which covers only a relatively small fraction of the worst-case circle. Therefore, only a small fraction of the people inside that circle would actually be affected by the cloud. Furthermore, since the endpoint of the cloud is much lower than the fatal toxic concentration, the number of fatalities resulting from the release would be smaller yet. For flammable gas scenarios, the blast effects would likely be felt in all directions from the source, so all people inside the circle could feel its effects. However, the endpoint for worst-case blast effects is 1 psi over pressure, which is also far below the level that would cause fatalities. Finally, the population count assumes all persons remain in place within the circle and are fully exposed for the time necessary to generate an effect; in an actual emergency, people shelter-in-place or evacuate and do not receive sufficient exposure to generate any ill effects.

- **How to cause a worst-case scenario.** Although the OCA indicates the scenario used to generate a release (e.g. vessel failure or hose rupture), the OCA contains no information describing how to make a worst-case or alternative scenario actually happen. OCA scenarios, and particularly worst-case scenarios, could generally only occur under a combination of very unusual conditions. Virtually no single event, such as detonation of a chemical explosive, could initiate a worst-case scenario. Studies of severe chemical plant accidents have shown that such accidents have usually resulted from the confluence of multiple abnormal events or conditions in process or management systems (*Loss Prevention in the Process Industries*, Second Edition, F.P. Lees, Butterworth-Heinemann Publishing, 1996, pp 2/2-3). In Bhopal, four separate safety systems, any one of which would have prevented the accident, had been disabled prior to the accident, and a fifth failed to operate properly (*Safety in the Chemical Industry, Lessons from Major Disasters*, E.A. Stallworthy and O.P. Kharbanda, G.P. Publishing, Columbia, MD, 1988, pp 99-100).

Investigations also show that single explosions rarely initiate worst-case toxic or flammable gas accidents. When viewed from a common-sense standpoint, the reason becomes clear. An explosion at a chemical plant often immediately starts a fire. If the initiating explosion also results in the release of a toxic substance, the ensuing fire will generally either partially or completely combust the toxin, reducing its toxicity while simultaneously dispersing it upward and away from surrounding populations. In the case of flammable materials, an initiating explosion will generally immediately ignite any flammable material released, causing a large fire, but actually preventing a much more severe vapor cloud explosion (the worst-case flammable accident). Vapor cloud explosions require that an extended release of flammable gas occur into a confined area prior to gas ignition. If the gas is immediately ignited upon release, a fire occurs, but not an explosion. Numerous experimental programs devoted to the study of vapor cloud explosions have shown that such explosions are difficult to reproduce, even under carefully controlled conditions (Guidelines for Evaluating the Characteristics of Vapor Cloud Explosions, Flash Fires, and BLEVEs, CCPS/AIChE, 1994, pp 70-75). Even accidental explosions at facilities that store, transport, and manufacture explosives have usually resulted in little damage (Stallworthy and Kharbanda, pp 67-68).

Finally, even if someone were able to trigger a vessel failure for example, the worst-case scenario assumes that all of the substance is released and becomes airborne within 10 minutes under extreme weather conditions. No one has any control over the weather and there is no guarantee that the vessel could be failed such that all of the chemical contained therein would be released quickly enough.

- **The specific location of toxic or flammable substances**. OCA information contains no information on site layout in general or the specific location of tanks, pipes, or vessels that contain toxic or flammable materials.

Other significant data which are not found in OCA information include:

- The location and design of release mitigation systems;
- Operating procedures for toxic or flammable material processes or release mitigation systems, or their set points or operating parameters;
- Design or construction information for any process equipment;
- Actual or prevailing meteorological conditions;

- Site security features or plans; and
- Site staffing plan or operations schedules

**11.     What OCA information is available to the public right now?**

While CSISSFRRA does not give the public a right of access to the EPA OCA information database prior to the promulgation of regulations, it does provide several mechanisms by which the public may nevertheless get access to limited amounts of OCA information before the regulations are issued; these mechanisms are described in detail in **Appendix B.**

**12.     What OCA data elements (i.e., OCA data *not* in RMP or EPA database format) are available to the public now?**

The public has access now to at least some of the OCA data elements in the RMP executive summary.  Facilities are required by the RMP regulations to include at least a brief description of their OCA in the executive summary of their RMPs; many facilities have included at least some of the OCA data elements in their summaries and those summaries are already posted on the Internet along with the rest of their RMPs (minus sections 2-5).  Facilities have wide latitude to decide how much OCA information to provide in their executive summary.  Check the **Sample RMP** in section 15 below.

**13.     What OCA "building block" data does the public have access to now?**

Other publicly available information can be gathered and analyzed to provide information similar to some OCA data elements.  For example, an important component of the worst-case release scenario is the quantity of the toxic or flammable substance in the largest vessel.  The chemical name and quantity of that chemical on-site is available in the registration section of the RMP for each process.  However, for a large facility with several listed substances and/or covered processes, the public won't know which chemical or process was considered for the potential worst-case or alternative release or how much of the chemical was expected to be released.  Also, the registration information identifies the quantity of the chemical in the process; this may or may not be the same as the quantity stored in the largest vessel.  Storage quantity is also available through TRI and other databases, but those data sources have the problems described above and more (e.g., TRI reports the total amount of a chemical on-site, not in a process)

If the quantity of the regulated substance in the registration information, or as provided by TRI, are used, it is likely that the scenario would over-estimate the consequence distance since the maximum quantity in a process or on-site is being used in lieu of the maximum amount stored in a single vessel.  Also, the analysis would not be able to determine or account for the existence of passive mitigation devices.

For worst-case scenario, the public has access to an EPA calculation model which most facilities have used to perform their analysis; however, without key inputs (such as those described above), people may get different (and potentially more alarming) results.

A-14

For alternative scenarios, the facility has wide latitude in its choice of inputs, so without key inputs, a member of the public attempting to reproduce a facility's OCA will likely get different results.

The potentially affected population is a function of the consequence distance, so whatever errors are made in determining this distance will also affect the public's calculation of population affected; census and other data for determining the population of a given area are publicly available.

The prevention program portion of RMPs lists all the types of mitigation measures used by a facility in a process, but without OCA-specific data, the public won't know which measures were involved in calculation of consequence distance or population affected.

Facility siting information can easily be found in telephone directories (e.g. yellow pages). In particular, sources of the yellow pages on the Internet are also linked to a map that shows the exact location of facilities. Also, siting and product information may be found in financial reports that are available on-line from the Securities and Exchange Commission's EDGAR system.

**14. What Other Consequence Information is available to the Public?**

Other information that may provide the public with information about the actual or potential consequences of accidental chemical releases include:

1.    The 5-year accident history section of RMPs.

2.    The accident prevention program sections of RMPs (including the hazard assessment).

3.    The accidental release information program is an EPA database that provides detailed information on the consequences and causes of a selected number of accidental releases.

4.    The Emergency Release Notification System (ERNS) is another EPA database that provides emergency notification information on any release of which the National Response Center has been notified. This information is provided to the extent it is known at the time of the release.

**15.    Sample RMP** (fictitious)

Below is a sample of a complete RMP including OCA information (fictitious) and how it might appear as printed from RMP*Info:

Facility Name: General Pulp & Paper
EPA ID:        1000 0010 1922

# Section 1. Registration Information

**1.1 Source Identification:**      **Facility ID: 12345**

     **a. Facility Name:**          General Pulp & Paper

     **b. Parent Company #1 Name:**

     **c. Parent Company #2 Name:**

**1.2 EPA Facility Identifier:**          1000 0010 1922

**1.3 Other EPA Systems Facility ID:**    ORD004201977

**1.4 Dun and Bradstreet Numbers (DUNS):**

     **a. Facility DUNS:**          001201977

     **b. Parent Company #1 DUNS:**

     **c. Parent Company #2 DUNS:**

**1.5 Facility Location Address:**

     **a. Street 1:**    238 Frontage Road

     **b. Street 2:**

     **c. City:**    Odenton          **d. State:** MD      **e. Zip:**    21873 -

     **f. County:**    HOWARD

**Facility Latitude and Longitude:**

     **g. Lat. (deg min sec):**    391115.0      **h. Long. (deg min sec):**    -0765010.0

     **g. Lat. (decimal degs.):**    45.187500      **h. Long. (decimal degs.):**    -076.8350

     **i. Lat/Long Method:**    I1      Interpolation - Map

     **j. Lat/Long Description:**    PG      Plant Entrance (General)

**1.6 Owner or Operator:**

     **a. Name:**    General Pulp & Paper

     **b. Phone:**    (410) 777-1234

     **Mailing address:**

     **c. Street 1:**   P.O. Box 1234          **d. Street 2:**

     **e. City:**   Odenton        **f. State:** MD    **g. Zip:**    21873 -

**1.7 Name and title of person or position responsible for part 68 (RMP) implementation:**

     **a. Name of person:**          John Jones

     **b. Title of person or position:**      Plant Manager

**1.8 Emergency contact:**

     **a. Name:**              Mary Smith

     **b. Title:**                Chemical Engineer

     **c. Phone:**            (410) 875-2871

     **d. 24-hour phone:**      (410) 875-4000

     **e. Ext. or PIN**

     **a. Facility or Parent Company E-Mail Address:**

     **b. Facility Public Contact Phone:**

     **c. Facility or Parent Company WWW Homepage Address:**

**1.10 LEPC:**      Howard County LEPC

**1.11 Number of full time employees on site:**          538

**1.12 Covered by:**

     **a. OSHA PSM:**      Yes

     **b. EPCRA 302:**      Yes

     **c. CAA Title V:**      Yes      **Air operating permit ID:**      06-2251

**1.13 OSHA Star or Merit Ranking:**      No

**1.14 Last Safety Inspection (by an External Agency) Date:**      08/21/1998

1.15 Last Safety Inspection Performed by an External Agency:    OSHA

1.16 Will this RMP involve predictive filing?:    No

### Reporting Center and RMP*Maintain Fields

| | | |
|---|---|---|
| Submission Method: | RMP*Submit | |
| Submission Type: | F | |
| Receipt Date: | 06/22/1999 | |
| Postmark Date: | 06/18/1999 | |
| Completeness Check Date: | 07/10/1999 | |
| Error Report Date: | | |
| De-registration Date: | | |
| De-registration Effective Date: | | |
| Anniversary Date: | | |

Certification Received: Yes
CBI Substantiation Letter: No
CBI Unsanitized Version: No
Electronic Waiver Present: No
Attachments Received: No
Graphic File Received: No
RMP Complete: Yes
CBI Flag: No

# Section 1.17 Process(es)

**a. Process ID:** 89876        **Program Level**    3    Chlorine System

**b. NAICS Code**

32211        Pulp Mills

**c. Process Chemicals**

| c.1 Process Chemical (ID / Name) | c.2 CAS Nr. | c.3 Qty (lbs.) |
|---|---|---|
| 19507        Chlorine | 7782-50-5 | 600,000 |

**a. Process ID:** 89877        **Program Level**    3    Chlorine Dioxide

**b. NAICS Code**

32211        Pulp Mills

**c. Process Chemicals**

| c.1 Process Chemical (ID / Name) | c.2 CAS Nr. | c.3 Qty (lbs.) |
|---|---|---|
| 19508        Chlorine dioxide [Chlorine oxide (ClO2)] | 10049-04-4 | 35,000 |

# Section 2. Toxics: Worst Case

## Toxics: Worst Case ID: 87654

2.1 a. Chemical Name:    Chlorine

   b. Percent Weight of Chemical (if in a mixture):

2.2 Physical State:    Both gas and liquid

2.3 Model used:    DEGADIS

2.4 Scenario:    Toxic gas release

2.5 Quantity released:        180,000   **lbs**

2.6 Release rate:        18,000.0   **lbs/min**

2.7 Release duration:        10.0   **mins**

2.8 Wind speed:        1.5   **m/sec**

2.9 Atmospheric Stability Class:    F

2.10 Topography:    Rural

2.11 Distance to Endpoint:        10.60   **mi**

2.12 Estimated Residential population within distance to endpoint:        156,567

**2.13 Public receptors within distance to endpoint:**

|  |  |  |  |  |
|---|---|---|---|---|
| a. Schools: | Yes | d. Prisons/Correction facilities: |  | Yes |
| b. Residences: | Yes | e. Recreation areas: |  | Yes |
| c. Hospitals: | Yes | f. Major commercial, office or, industrial areas: |  | Yes |
| g. Other (Specify): |  |  |  |  |

**2.14 Environmental receptors within distance to endpoint:**

| | |
|---|---|
| a. National or state parks, forests, or monuments: | Yes |
| b. Officially designated wildlife sanctuaries, preserves, or refuges: | Yes |
| c. Federal wilderness areas: | No |
| d. Other (Specify): | |

**2.15 Passive mitigation considered:**

|  |  |  |  |
|---|---|---|---|
| a. Dikes: | No | d. Drains: | No |
| b. Enclosures: | No | e. Sumps: | No |
| c. Berms: | No | f. Other (Specify): | |

## Toxics: Worst Case ID: 87655

**2.1 a. Chemical Name:**   Chlorine dioxide  [Chlorine oxide (ClO2)]

    **b. Percent Weight of Chemical (if in a mixture):**

**2.2 Physical State:**   Gas

**2.3 Model used:**   DEGADIS

**2.4 Scenario:**   Toxic gas release

**2.5 Quantity released:**   19,860  lbs

**2.6 Release rate:**   1,986.0  lbs/min

**2.7 Release duration:**   10.0  mins

**2.8 Wind speed:**   1.5  m/sec

**2.9 Atmospheric Stability Class:**   F

**2.10 Topography:**   Rural

**2.11 Distance to Endpoint:**   7.20  mi

**2.12 Estimated Residential population within distance to endpoint:**   26,240

**2.13 Public receptors within distance to endpoint:**

|  |  |  |  |  |
|---|---|---|---|---|
| a. Schools: | Yes | d. Prisons/Correction facilities: |  | Yes |
| b. Residences: | Yes | e. Recreation areas: |  | Yes |
| c. Hospitals: | Yes | f. Major commercial, office or, industrial areas: |  | Yes |
| g. Other (Specify): |  |  |  |  |

**2.14 Environmental receptors within distance to endpoint:**

| | |
|---|---|
| a. National or state parks, forests, or monuments: | No |
| b. Officially designated wildlife sanctuaries, preserves, or refuges: | Yes |
| c. Federal wilderness areas: | No |
| d. Other (Specify): | |

|  |  |  |  |
|---|---|---|---|
| a. Dikes: | No | d. Drains: | No |
| b. Enclosures: | No | e. Sumps: | No |
| c. Berms: | No | f. Other (Specify): | |

# Section 3. Toxics: Alternative Release

## Toxics: Alternative Release ID: 98765

3.1 a. Chemical Name:     Chlorine

b. Percent Weight of Chemical (if in a mixture):

3.2 Physical State:     Both gas and liquid

3.3 Model:     DEGADIS

3.4 Scenario:     Pipe leak

3.5 Quantity released:     310 **lbs**

3.6 Release rate:     1550.0 **lbs/min**

3.7 Release duration:     0.2 **mins**

3.8 Wind speed:     3.0 **m/sec**

3.9 Atmospheric Stability Class:     D

3.10 Topography:     Rural

3.11 Distance to Endpoint:     1.80 **mi**

3.12 Estimated Residential population within distance to endpoint:     3,930

3.13 Public receptors within distance to endpoint:

| | | | |
|---|---|---|---|
| a. Schools: | Yes | d. Prisons/Correction facilities: | No |
| b. Residences: | Yes | e. Recreation areas: | Yes |
| c. Hospitals: | No | f. Major commercial, office, or industrial areas: | Yes |
| g. Other (Specify): | | | |

3.14 Environmental receptors within distance to endpoint:

| | |
|---|---|
| a. National or state parks, forests, or monuments: | No |
| b. Officially designated wildlife sanctuaries, preserves, or refuges: | No |
| c. Federal wilderness areas: | No |
| d. Other (Specify): | |

3.15 Passive mitigation considered:

| | | | |
|---|---|---|---|
| a. Dikes: | No | d. Drains: | No |
| b. Enclosures: | No | e. Sumps: | No |
| c. Berms: | No | f. Other (Specify): | |

3.16 Active mitigation considered:

| | | | |
|---|---|---|---|
| a. Sprinkler systems: | No | f. Flares: | No |
| b. Deluge system: | No | g. Scrubbers: | No |
| c. Water curtain: | No | h. Emergency shutdown systems: | Yes |
| d. Neutralization: | No | i. Other (Specify): | |
| e. Excess flow valve: | | | |

## Toxics: Alternative Release ID: 98766

3.1 a. Chemical Name:     Chlorine dioxide [Chlorine oxide (ClO2)]

b. Percent Weight of Chemical (if in a mixture):

3.2 Physical State:     Gas

3.3 Model:     DEGADIS

3.4 Scenario:     Pipe leak

3.5 Quantity released:     1,510 **lbs**

3.6 Release rate:     80.0 **lbs/min**

3.7 Release duration:     20.0 **mins**

**3.8 Wind speed:** 3.0 m/sec

**3.9 Atmospheric Stability Class:** D

**3.10 Topography:** Rural

**3.11 Distance to Endpoint:** 2.13 mi

**3.12 Estimated Residential population within distance to endpoint:** 4,660

**3.13 Public receptors within distance to endpoint:**

| | | | |
|---|---|---|---|
| a. Schools: | Yes | d. Prisons/Correction facilities: | No |
| b. Residences: | Yes | e. Recreation areas: | Yes |
| c. Hospitals: | Yes | f. Major commercial, office, or industrial areas: | Yes |
| g. Other (Specify): | | | |

**3.14 Environmental receptors within distance to endpoint:**

| | |
|---|---|
| a. National or state parks, forests, or monuments: | No |
| b. Officially designated wildlife sanctuaries, preserves, or refuges: | No |
| c. Federal wilderness areas: | No |
| d. Other (Specify): | |

**3.15 Passive mitigation considered:**

| | | | |
|---|---|---|---|
| a. Dikes: | No | d. Drains: | Yes |
| b. Enclosures: | No | e. Sumps: | Yes |
| c. Berms: | No | f. Other (Specify): | |

**3.16 Active mitigation considered:**

| | | | |
|---|---|---|---|
| a. Sprinkler systems: | No | f. Flares: | No |
| b. Deluge system: | No | g. Scrubbers: | No |
| c. Water curtain: | No | h. Emergency shutdown systems: | No |
| d. Neutralization: | No | i. Other (Specify): | |
| e. Excess flow valve: | No | | |

# Section 4. Flammables: Worst Case --- No Data To Report

# Section 5. Flammables: Alternative Release --- No Data To Report

# Section 6. Accident History

**Accident History ID:** 3567

**6.1 Date of accident:** 12/03/1996     **6.2 Time accident began(HHMM):** 1330

**6.3 NAICS Code of process involved:** 32211

**6.4 Release duration:** 000 Hours (HHH)     20 Minutes (MM)

**6.5 Chemical(s):**

| a. Chemical Name | CAS Number | b. Quantity Released (lbs) | c. % Weight |
|---|---|---|---|
| Chlorine | 7782-50-5 | 1 | |

**6.6 Release event:**          **6.7 Release source:**

| | | | | | |
|---|---|---|---|---|---|
| a. Gas release: | Yes | a. Storage vessel: | No | e. Valve: | No |
| b. Liquid spill/evaporation: | No | b. Piping: | No | f. Pump: | No |
| c. Fire: | No | c. Process vessel: | No | g. Joint: | No |

| d. Explosion: | No | d. Transfer hose: | Yes | h. Other (Specify): |

**6.8 Weather conditions at time of event (if known):**

    a. Wind speed:                 **Units:** meters/second       **Direction:**

    b. Temperature:               Degrees Fahrenheit

    c. Atmospheric Stability Class:

    d. Precipitation present:    No

    e. Unknown weather conditions:    Yes

**6.9 On-site impacts:**

| | Employees or contractors: | Public responders: | Public: |
|---|---|---|---|
| a. Deaths | 0 | 0 | 0 |
| b. Injuries | 1 | 0 | 0 |
| c. Property damage ($): | 2,500 | | |

**6.10 Known Off-site impacts:**

| | | | |
|---|---|---|---|
| a. Deaths: | 0 | d. Evacuated: | 0 |
| b. Hospitalization: | 0 | e. Sheltered-in-place: | 0 |
| c. Other medical treatments: | 0 | f. Property Damage ($): | 0 |

    g. Environmental damage:

        1. Fish or Animal Kills:         No

        2. Tree, lawn, shrub, or crop damage:    No

        3. Water contamination:         No

        4. Soil contamination:          No

        5. Other (specify):

**6.11 Initiating event:**      a      Equipment Failure

**6.12 Contributing factors:**

| | | | | |
|---|---|---|---|---|
| a. Equipment failure: | Yes | g. Maintenance activity/inactivity: | No |
| b. Human error: | Yes | h. Process design failure: | No |
| c. Improper procedures: | No | i. Unsuitable equipment: | No |
| d. Overpressurization: | No | j. Unusual weather condition: | No |
| e. Upset condition: | No | k. Management error: | No |
| f. By-pass condition: | No | l. Other (Specify): | |

**6.13 Offsite responders notified:**      Notified and Responded

**6.14 Changes introduced as a result of the accident:**

| | | | |
|---|---|---|---|
| a. Improved or upgraded equipment: | Yes | g. Revised emergency response plan: | No |
| b. Revised maintenance: | No | h. Changed process: | No |
| c. Revised training: | No | i. Reduced inventory: | No |
| d. Revised operating procedures: | Yes | j. None: | No |
| e. New process controls: | No | k. Other(Specify): | |
| f. New mitigation systems: | No | | |

# Section 7. Prevention Program 3

**Process ID:** <u>**89876**</u>  Chlorine System

**Prevention Program ID:** **8049**

**Prevention Program Description:**

Chlorine Process

**7.1 NAICS Code**  32211

**7.2 Chemicals**  **Chemical Name**
Chlorine

**7.3 Date on which the safety information was last reviewed or revised:**  06/11/1998

**7.4  Process Hazard Analysis (PHA):**

a. Date of last PHA or PHA update:  03/05/1999

b. The technique used:

| | | | |
|---|---|---|---|
| What If: | No | Failure Mode and Effects Analysis: | No |
| Checklist: | No | Fault Tree Analysis: | No |
| What If/Checklist: | No | HAZOP: | Yes |
| Other (Specify): | | | |

c. Expected or actual date of completion of all changes  from last PHA or PHA update:  06/11/1999

d. Major hazards identified:

| | | | |
|---|---|---|---|
| Toxic release: | Yes | Contamination: | Yes |
| Fire: | Yes | Equipment failure: | Yes |
| Explosion: | Yes | Loss of cooling, heating, electricity, instrument air: | Yes |
| Runaway reaction: | No | Earthquake: | Yes |
| Polymerization: | No | Floods (flood plain): | Yes |
| Overpressurization: | Yes | Tornado: | No |
| Corrosion: | Yes | Hurricanes: | No |
| Overfilling: | Yes | Other (Specify): | |

e. Process controls in use:

| | | | |
|---|---|---|---|
| Vents: | Yes | Emergency air supply: | No |
| Relief valves: | Yes | Emergency power: | Yes |
| Check valves: | Yes | Backup pump: | No |
| Scrubbers: | Yes | Grounding equipment: | No |
| Flares: | No | Inhibitor addition: | No |
| Manual shutoffs: | Yes | Rupture disks: | Yes |
| Automatic shutoffs: | Yes | Excess flow device: | Yes |
| Interlocks: | Yes | Quench system: | No |
| Alarms and procedures: | No | Purge system: | Yes |
| Keyed bypass: | No | None: | No |
| Other (Specify): | | | |

f. Mitigation systems in use:

| | | | |
|---|---|---|---|
| Sprinkler system: | No | Water curtain: | No |
| Dikes: | No | Enclosure: | Yes |
| Fire walls: | No | Neutralization: | Yes |
| Blast walls: | No | None: | No |

Deluge system:     No          Other (Specify):

g. Monitoring/detection systems in use:

| | | | |
|---|---|---|---|
| Process area detectors: | Yes | **None:** | No |
| Perimeter monitors: | No | **Other (Specify):** | Video Surveillance |

h. Changes since last PHA or PHA update:

| | | | |
|---|---|---|---|
| Reduction in chemical inventory: | No | **Installation of perimeter monitoring systems:** | No |
| Increase in chemical inventory: | No | **Installation of mitigation systems:** | No |
| Change process parameters: | No | **None recommended:** | Yes |
| Installation of process controls: | No | **None:** | No |
| Installation of process detection systems: | No | **Other (Specify):** | |

7.5 Date of most recent review or revision of operating procedures:     05/01/1999

7.6 Training:

  a. The date of the most recent review or revision of training programs:     05/01/1999

  b. The type of training provided:

    **Classroom:** Yes    **On the job:** Yes    **Other (Specify):**

  c. The type of competency testing used:

| | | | |
|---|---|---|---|
| **Written test:** | Yes | **Observation:** | Yes |
| **Oral test:** | No | **Demonstration:** | Yes |
| **Other (Specify):** | | | |

7.7 Maintenance:

  a. The date of the most recent review or revision of maintenance procedures:     05/01/1999

  b. The date of the most recent equipment inspection or test:     06/11/1999

  c. Equipment most recently inspected or tested :     piping

7.8 Management of change:

  a. The date of the most recent change that triggered management of change procedures:     05/24/1999

  b. The date of the most recent review or revision of management of change procedures:     3/18/1999

7.9 The date of the most recent pre-startup review:     08/22/1998

7.10 Compliance audits:

  a. The date of the most recent compliance audit:     06/11/1999

  b. Expected date of completion of all changes resulting from the compliance audit:     06/18/1999

7.11 Incident investigation:

  a. The date of the most recent incident investigation (if any):     12/04/1996

  b. Expected or actual date of completion of all changes resulting from the investigation:     03/15/1997

7.12 The date of the most recent review or revision of employee participation plans:     03/11/1999

7.13 The date of the most recent review or revision of hot work permit procedures:     04/13/1999

7.14 The date of the most recent review or revision of contractor safety procedures:     04/18/1999

7.15 The date of the most recent evaluation of contractor safety performance:     04/04/1999

**Process ID:** __89877__    Chlorine Dioxide

**Prevention Program ID:**    **8050**

**Prevention Program Description:**
Chlorine Dioxide System

**7.1 NAICS Code**    32211

**7.2 Chemicals**      **Chemical Name**
Chlorine dioxide [Chlorine oxide (ClO2)]

**7.3 Date on which the safety information was last reviewed or revised:**      06/30/1997

**7.4 Process Hazard Analysis (PHA):**

     a. Date of last PHA or PHA update:      02/16/1999

     b. The technique used:

| | | | |
|---|---|---|---|
| **What If:** | No | **Failure Mode and Effects Analysis:** | No |
| **Checklist:** | No | **Fault Tree Analysis:** | No |
| **What If/Checklist:** | No | **HAZOP:** | Yes |
| **Other (Specify):** | | | |

     c. Expected or actual date of completion of all changes from last PHA or PHA update:      05/18/1999

     d. Major hazards identified:

| | | | |
|---|---|---|---|
| **Toxic release:** | Yes | **Contamination:** | Yes |
| **Fire:** | Yes | **Equipment failure:** | Yes |
| **Explosion:** | Yes | **Loss of cooling, heating, electricity, instrument air:** | Yes |
| **Runaway reaction:** | No | **Earthquake:** | Yes |
| **Polymerization:** | No | **Floods (flood plain):** | Yes |
| **Overpressurization:** | Yes | **Tornado:** | No |
| **Corrosion:** | Yes | **Hurricanes:** | No |
| **Overfilling:** | Yes | **Other (Specify):** | |

     e. Process controls in use:

| | | | |
|---|---|---|---|
| **Vents:** | Yes | **Emergency air supply:** | No |
| **Relief valves:** | Yes | **Emergency power:** | Yes |
| **Check valves:** | Yes | **Backup pump:** | No |
| **Scrubbers:** | Yes | **Grounding equipment:** | No |
| **Flares:** | No | **Inhibitor addition:** | No |
| **Manual shutoffs:** | Yes | **Rupture disks:** | Yes |
| **Automatic shutoffs:** | Yes | **Excess flow device:** | Yes |
| **Interlocks:** | Yes | **Quench system:** | No |
| **Alarms and procedures:** | Yes | **Purge system:** | Yes |
| **Other (Specify):** | | | |

     f. Mitigation systems in use:

| | | | |
|---|---|---|---|
| **Sprinkler system:** | No | **Water curtain:** | No |
| **Dikes:** | No | **Enclosure:** | Yes |
| **Fire walls:** | No | **Neutralization:** | Yes |
| **Blast walls:** | No | **None:** | No |
| **Deluge system:** | No | **Other (Specify):** | |

     g. Monitoring/detection systems in use:

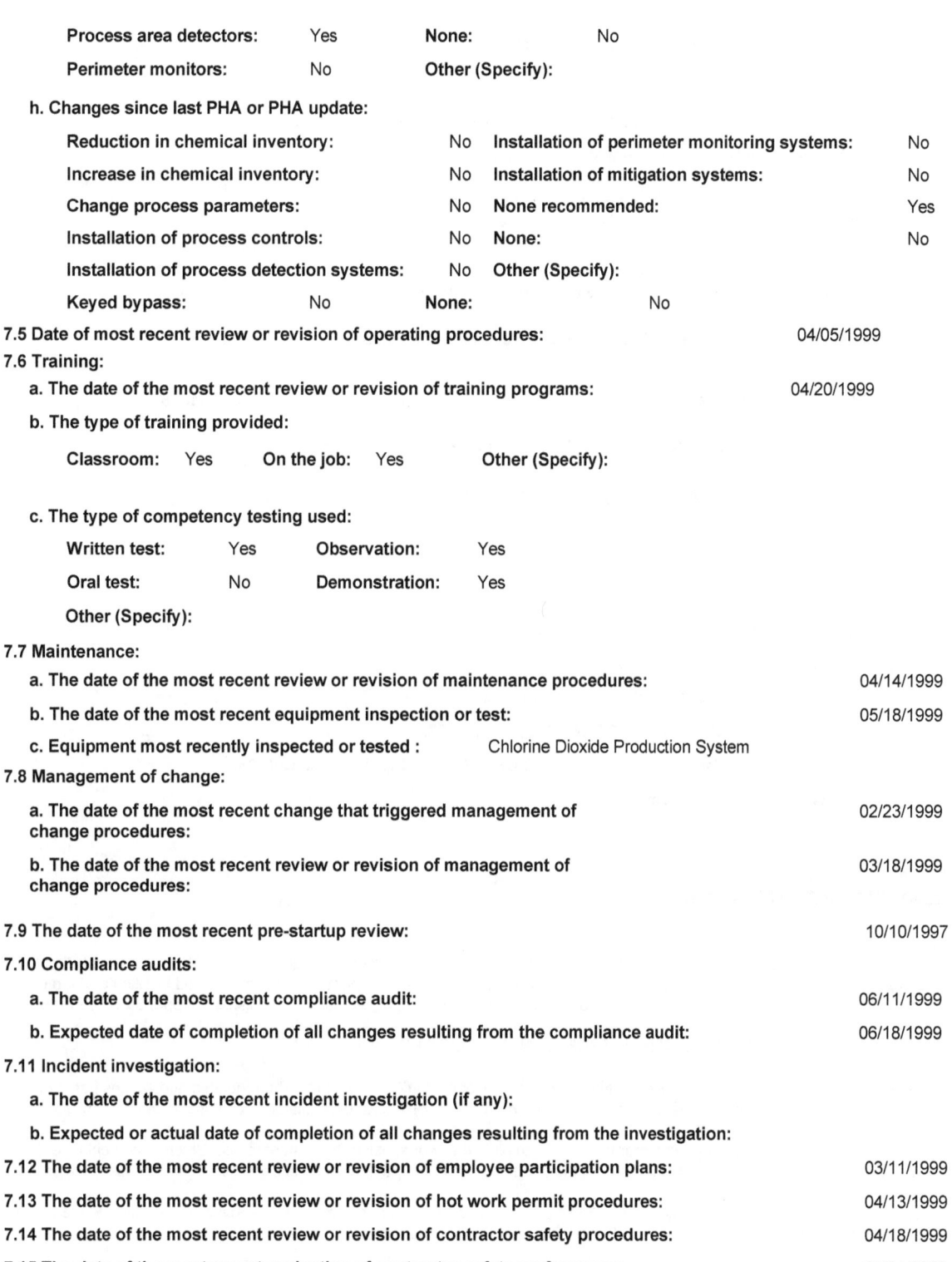

| Process area detectors: | Yes | **None:** | No |
| Perimeter monitors: | No | **Other (Specify):** | |

h. Changes since last PHA or PHA update:

| Reduction in chemical inventory: | No | **Installation of perimeter monitoring systems:** | No |
| Increase in chemical inventory: | No | **Installation of mitigation systems:** | No |
| Change process parameters: | No | **None recommended:** | Yes |
| Installation of process controls: | No | **None:** | No |
| Installation of process detection systems: | No | **Other (Specify):** | |
| Keyed bypass: | No | **None:** | No |

**7.5 Date of most recent review or revision of operating procedures:**      04/05/1999

**7.6 Training:**

  a. The date of the most recent review or revision of training programs:      04/20/1999

  b. The type of training provided:

    **Classroom:**   Yes    **On the job:**   Yes      **Other (Specify):**

  c. The type of competency testing used:

| **Written test:** | Yes | **Observation:** | Yes |
| **Oral test:** | No | **Demonstration:** | Yes |

    **Other (Specify):**

**7.7 Maintenance:**

  a. The date of the most recent review or revision of maintenance procedures:      04/14/1999

  b. The date of the most recent equipment inspection or test:      05/18/1999

  c. Equipment most recently inspected or tested :      Chlorine Dioxide Production System

**7.8 Management of change:**

  a. The date of the most recent change that triggered management of change procedures:      02/23/1999

  b. The date of the most recent review or revision of management of change procedures:      03/18/1999

**7.9 The date of the most recent pre-startup review:**      10/10/1997

**7.10 Compliance audits:**

  a. The date of the most recent compliance audit:      06/11/1999

  b. Expected date of completion of all changes resulting from the compliance audit:      06/18/1999

**7.11 Incident investigation:**

  a. The date of the most recent incident investigation (if any):

  b. Expected or actual date of completion of all changes resulting from the investigation:

**7.12 The date of the most recent review or revision of employee participation plans:**      03/11/1999

**7.13 The date of the most recent review or revision of hot work permit procedures:**      04/13/1999

**7.14 The date of the most recent review or revision of contractor safety procedures:**      04/18/1999

**7.15 The date of the most recent evaluation of contractor safety performance:**      04/04/1999

# Section 8. Prevention Program 2 --- No Data To Report

# Section 9. Emergency Response

**9.1 Written Emergency Response (ER) Plan:**

    a. Is facility included in written community emergency response plan?     Yes

    b. Does facility have its own written emergency response plan?     Yes

**9.2 Does facility's ER plan include specific actions to be taken in response to accidental releases of regulated substance(s)?**     Yes

**9.3 Does facility's ER plan include procedures for informing the public and local agencies responding to accidental releases?**     Yes

**9.4 Does facility's ER plan include information on emergency heath care?**     Yes

**9.5 Date of most recent review or update of facility's ER plan:**     11/30/1998

**9.6 Date of most recent ER training for facility's employees:**     02/06/1999

**9.7 Local agency with which facility's ER plan or response activities are coordinated:**

    a. Name of agency:     Howard County Fire Department

    b. Telephone number:     (410)-321-7654

**9.8 Subject to:**

    a. OSHA  Regulations at 29 CFR 1910.38:     Yes

    b. OSHA Regulations at 29 CFR 1910.120:     Yes

    c. Clean Water Act Regulations at 40 CFR 112:     Yes

    d. RCRA Regulations at 40 CFR 264, 265, and 279.52:     Yes

    e. OPA-90 Regulations at 40 CFR 112, 33 CFR 154, 49 CFR 194, or 30 CFR 254:     Yes

    f. State EPCRA Rules/Law:     No

# Executive Summary

This Risk Management Plan (RMP) is submitted to the U.S. Environmental Protection Agency (U.S. EPA) for General Pulp & Paper in accordance with the requirements of Section 112(r) of the Clean Air Act Amendments of 1990 as codified in Title 40 of the Code of Federal Regulations (CFR) Part 68. General Pulp & Paper handles two regulated substances listed in Appendix A of Part 68.

1.1 STATIONARY SOURCE & REGULATED SUBSTANCES HANDLED

General Pulp & Paper owns and operates a pulp and paper mill located in Odenton, Md. The regulated substances handled by this facility are chlorine and chlorine dioxide, both of which are on the U.S. EPA's list of regulated toxic substances for CAA section 112(r).

The Odenton plant produces pulp and paper from wood chips and sawdust using the Kraft process. Chlorine and chlorine dioxide are used in a bleaching process to remove lignin from the fibers and to whiten pulp; chlorine is also used to treat process water.

Liquid chlorine is stored in rail cars and storage tanks prior to use and fed to a vaporizer. Gaseous chlorine is then fed to the process. The maximum quantity of chlorine that stored at this facility is 600,000 pounds.

Chlorine dioxide is generated on site by a process which uses sodium chlorate, methanol, and sulfuric acid as a raw materials. These raw materials are not regulated under section 112(r). The chlorine dioxide produced in the process is absorbed into water and then stored as a dilute aqueous solution (10 g/l). The maximum quantity of chlorine dioxide stored at this facility is 35,000 pounds.

1.2 ACCIDENTAL RELEASE PREVENTION & EMERGENCY RESPONSE PROGRAMS

General Pulp & Paper prevents chemical accidents using an integrated process safety management system. The plant uses several management systems and follows applicable industry and national standards to meet this goal.

General Pulp & Paper's chlorine and chlorine dioxide processes are covered by the OSHA Process Safety Management (PSM) standard (29 CFR 1910.119). General Pulp & Paper adheres strictly to the PSM standard and focuses many of its safety efforts around PSM. The PSM program requires General Pulp & Paper to take specific efforts to identify and mitigate process hazards and prevent accidents. The elements of the PSM program are very similar to the accident prevention elements in the EPA risk management program, which General Pulp and Paper also fully implements.

Although an accidental chemical release is unlikely, General Pulp & Paper prepares for releases and other emergencies. The plant has developed and implemented a written response plan which is discussed is detail in Section 1.5. General Pulp & Paper employees routinely practice responding to simulated releases and emergencies, and coordinate with community responders such as the Odenton Fire Department.

## 1.3 WORST-CASE & ALTERNATIVE RELEASE SCENARIOS

General Pulp & Paper has constructed a worse-case release scenario and alternate (i.e. more credible) release scenario for each regulated chemical.

### CHLORINE: WORST-CASE SCENARIO

The failure of the largest storage tank (i.e. railcar) when filled to the greatest amount allowed would release 180,000 pounds of chlorine. Since the contents of the railcar are under pressure, the release is assumed to be a liquid jet that volatilizes to gas upon release from the tank. The entire contents of the railcar are assumed to release at a constant rate over a ten minute period.

### CHLORINE DIOXIDE: WORST-CASE SCENARIO

The failure of our largest chlorine dioxide solution storage tanks would release 238,000 gallons of chlorine dioxide solution, or 19,856 pounds of chlorine dioxide. Company policy limits the maximum filling capacity of the large chlorine dioxide storage tanks to 90%; the 238,000 gallon figure is 90% of the physical capacity of the tank. It is assumed that the entire contents of the tank are released and instantaneously form a pool 1 cm deep. The chlorine dioxide volatilization rate from the pool is calculated according to a model based on an evaporative pool model.

### CHLORINE: ALTERNATE SCENARIO

A 1" pipe conveys liquid chlorine to the water treatment plant from the chlorine expansion tank. This pipe could be ruptured by a vehicle (e.g. forklift) striking the pipe bridge which contains the chlorine pipe. This would release 310 pounds of liquid chlorine that is assumed to vaporize instantly. The release is estimated to take twelve seconds.

### CHLORINE DIOXIDE: ALTERNATE SCENARIO

A fiberglass pipe which conveys chlorine dioxide from the large storage tanks to the bleach plant is assumed to be damaged by mechanical impact during a pump replacement or other maintenance work. A 3" diameter hole is made in the pipe and chlorine dioxide solution is released. The motive force is the gravity head of the tank; it is assumed that the pump is shut off immediately during the evacuation of the area. The release continues for twenty minutes until a response crew can enter the required protective equipment and shut off the release. A drain in the vicinity of the pipe is assumed to be able to capture 2 gallons per second of the spill; this is directed to a gas-tight sump where the spilled material can be collected and treated.

## 1.4 FIVE YEAR ACCIDENT HISTORY

General Pulp & Paper has had one release of a regulated material that resulted in an injury in the last five years. On December 3, 1996, an employee was injured when exposed to chlorine leaking from a hose. There have been no releases of regulated materials which have resulted in deaths, significant property damage, or any known offsite deaths, injuries, evacuations, sheltering in place, property damage, or environmental damage in the last five years.

## 1.5 EMERGENCY RESPONSE PROGRAM

In addition to the prevention program, General Pulp & Paper has developed and implemented a written emergency response plan to effectively respond to accidental chemical releases. This plan identifies roles for plant personnel in the event of a number of different scenarios. The plan includes specific tasks for key personnel during responses, emergency plant shutdown procedures, steps to contain and handle releases of specific materials, specific information on how to contact community response agencies and the public, and information on training employees and community responders in safe response techniques. General Pulp & Paper trains regularly on its emergency plan. This training includes plant employees, members of General Pulp & Paper's response team, and community responders. Training exercises are evaluated, and the plan is updated when deficiencies are identified.

General Pulp & Paper maintains an emergency response team that is trained to respond to many different types of emergencies. The team is made up of workers from different shifts and is always ready to respond. The team regularly conducts response drills, often including community responders.

## 1.6 PLANNED CHANGES TO IMPROVE SAFETY

General Pulp & Paper has identified no major unresolved process hazards in the chlorine or chlorine dioxide systems. No major revisions to those processes are currently planned. However, General Pulp & Paper follows a policy of continuous process safety improvement.

# APPENDIX B

# DISCUSSION OF THE APPLICABLE PROVISIONS OF THE CHEMICAL SAFETY INFORMATION, SITE SECURITY, AND FUELS REGULATORY RELIEF ACT (CSISSFRRA)

# APPENDIX B

# DISCUSSION OF THE APPLICABLE PROVISIONS OF THE CHEMICAL SAFETY INFORMATION, SITE SECURITY, AND FUELS REGULATORY RELIEF ACT (CSISSFRRA)

**1.      What is CSISSFRRA, and how does it relate to this assessment?**

In August of 1999, Congress passed the Chemical Safety Information, Site Security and Fuels Regulatory Relief Act (CSISSFRRA) to address concerns about potential Internet posting of a database containing information on the consequences of hypothetical chemical accidents.[1] Under a regulatory program required by the Clean Air Act (CAA), facilities handling certain very hazardous substances must conduct analyses of the off-site consequences of such hypothetical accidents and report the results (off-site consequence analysis information or OCA information - see Appendix A) in a plan submitted to the U.S. Environmental Protection Agency (EPA). CSISSFRRA temporarily exempts OCA information from public disclosure under the CAA and the Freedom of Information Act (FOIA).

CSISSFRRA also requires the President to assess the increased risk of terrorist and other criminal activity associated with the posting of OCA information on the Internet; and the incentives created by public disclosure of OCA information to reduce the risk of accidental chemical releases. Based on the assessments, the President is to issue regulations governing the distribution of OCA information in a manner that, in the opinion of the President, minimizes the likelihood of accidental releases and any increased risk of terrorist activity associated with Internet posting of OCA information and the likelihood of harm to public health and welfare. The President has delegated to the Department of Justice (DOJ) the responsibility of assessing the increased risk of terrorist and criminal activity and to EPA the responsibility of assessing the incentives for reduction in chemical accidents created by public disclosure of OCA information. On January 27, 2000, the President provided joint delegation to DOJ and EPA to promulgate the regulations, after review and approval by the Office of Management and Budget.

As noted above, CSISSFRRA exempts "off-site consequence analysis information" from FOIA for at least one year while the President assesses the criminal risks of posting the information on the Internet and the chemical safety benefits of providing public access to the information and then issues regulations governing distribution of the information based on the assessments (section 112(r)(7)(H)(ii)). CSISSFRRA defines "off-site consequence analysis information" (OCA information) as the OCA sections of any RMP submitted to EPA and any electronic database EPA creates from those sections (section 112(r)(7)(H)(I)(III)). It expressly

---

[1] CSISSFRRA also contains provisions which prohibit EPA from regulating flammable substances under the Risk Management Program when those substances are used as fuel or held for sale as fuel at a retail facility. The fuel provisions of CSISSFRRA are not discussed in this Appendix.

excludes an RMP's Executive Summary, which is required to include at least a brief description of the submitting facility's OCA.

## 2.    How can the public gain access to OCA information under CSISSFRRA?

CSISSFRRA provides the public with other means of access to the data reported in the OCA sections of RMPs, and even to the OCA sections themselves, before and/or after the federal government conducts its assessments and rulemaking.  As noted above, RMP Executive Summaries are not covered by its restrictions, and facilities are required to provide at least a brief description of their OCAs in their Executive Summaries.  The summaries are already available on the Internet through several web sites.  A random sampling of the summaries indicates that the amount of OCA information reported varies from facility to facility; some facilities provided nearly complete information while others provided little.  (EPA's rule does not define a "brief description," leaving facilities to make reasonable decisions as to what information to include.)  In addition, CSISSFRRA requires virtually all covered facilities to conduct a public meeting or post a public notice by February 5, 2000, that summarizes their OCA information (CSISSFRRA section 4).  To date, the Federal Bureau of Investigation has received notification from about 5,000 facilities that they have complied with this requirement.

CSISSFRRA also does not prevent facilities from releasing their OCA information to the public without restriction, and once a facility has so released its OCA information, covered persons may do so as well (section 112(r)(7)(H)(v)(III)).  To date, EPA has received notification from over 900 facilities that they have released their OCA information without restriction. CSISSFRRA further provides that states which collect data on off-site consequences, even data identical in content and format to OCA information, are not precluded from releasing it to the public (section 112(r)(7)(H)(x)(II)).  Several states have in place state laws requiring the collection of similar or identical data.

CSISSFRRA guarantees public access to OCA information itself (i.e., the OCA sections of RMPs and EPA's database created from those sections) in several ways.  First, it requires EPA to provide the public with OCA information without information concerning the identity and location of the facilities reporting the information (section 112(r)(7)(H)(iv)).  EPA is consulting with other federal agencies and stakeholders to implement this provision.  Second, CSISSFRRA requires EPA, in consultation with DOJ and other agencies, to establish a "read-only information technology system" that "provides for the availability to the public of [OCA information] by means of a central data base under the control of the Federal Government that contains information that users may read, but that provides no means by which an electronic or mechanical copy of the information may be made" (section 112(r)(7)(H)(viii)).  EPA is working with other federal agencies to identify the best methods for development of this read-only system.   Third, CSISSFRRA requires EPA, in consultation with DOJ, to make OCA information available to "qualified researchers" by means of a system that does not allow researchers who receive the information to disseminate it (section 112(r)(7)(H)(vii)).  Finally, CSISSFRRA provides that, at a minimum, the regulations based on the assessment must "allow access by any member of the public to paper copies of [OCA] information for a limited number of stationary sources located anywhere in the United States, without any geographical restriction." In short, any member of the public will be able to have access to paper copies of OCA information for at least some number of facilities.

### 3. Who are "covered persons" and how does CSISSFRRA affect them?

Both before and after the regulations are issued, CSISSFRRA guarantees "covered persons" access to OCA information for their "official use" (see section 112(r)(7)(H)(iv) and (ii)(cc)-(ee)).

The SERC/LEPC category includes members of 50 State Emergency Response Commissions and about 3,400 Local Emergency Planning Committees created under EPCRA. Members of these commissions and committees can include members of public, the media, and industry, as well as representatives of emergency responders such as fire and police departments (EPCRA section 301(c)). Considering that covered persons include all of the entities above, there are potentially well over 1 million covered persons.

> **"Covered persons"** include:
> - Federal government officers and employees and their contractors
> - State government officers and employees and their contractors
> - Local government officers and employees and their contractors
> - SERC and LEPC members and their contractors
> - State and local police
> - Paid and volunteer firefighters
> - Other emergency responders

Covered persons are guaranteed access to OCA information (in either RMP or database format) for any or all covered facilities for "official use" both now and in the future (i.e., under the regulations); members of the public have no right of access prior to the regulations, but are guaranteed access to at least paper copies of the OCA sections of RMPs for a limited number of facilities under the regulations and to a read-only database of OCA information that EPA is to establish in consultation with other federal agencies.

While CSISSFRRA guarantees covered persons access to OCA information, it prohibits them from disclosing the information to the public except as authorized by the statute or the regulations issued under it (section 112(r)(7)(H)(v)). Any covered person who violates the prohibition is subject to criminal penalties of up to $1,000,000 per year. At the same time, CSISSFRRA states that it "does not restrict the dissemination of off-site consequence analysis information by any covered person in any manner or form except in the form of a [RMP] or an electronic data base created by [EPA] from off-site consequence analysis information" (section 112(r)(7)(H)(xii)(II)). CSISSFRRA's prohibition on public disclosure is thus narrow. It applies to the OCA sections of the RMP (sections 2-5 of the RMP form) and any database created by EPA from those sections, but it does not apply to the information reported in those sections when provided in a different format, or to the information provided in Executive Summaries. Covered persons are consequently allowed to communicate the information in the OCA sections of RMPs to the public so long as they do so in a way that does not replicate those sections of the RMP or EPA's database.

This page intentionally left blank.

# APPENDIX C

# USES OF RIGHT-TO-KNOW INFORMATION

# APPENDIX C

## USES OF RIGHT-TO-KNOW INFORMATION

This appendix summarizes over 40 documented cases in which right-to-know information was used to improve conditions for communities. The first section highlights uses of EPCRA (Emergency Planning and Community Right-to-Know Act) data. Most of these uses rely on Toxics Release Inventory (TRI) data, which are available electronically from a central source, but there are some uses of other EPCRA data as well. The second section describes uses of information from environmental and other programs outside of EPCRA to reduce risk in some form.

These uses of right-to-know information shed light on OCA information – on how and how much it would likely be used under various disclosure schemes. For example, usage of easily available TRI data seems to be much higher than usage of other EPCRA data, which are more difficult to obtain. In addition, there is a wide variety of users and of uses of right-to-know information. Most cases involve multiple segments of the public, with the primary actors ranging from community, public interest, and environmental organizations, to news media, government, industry, unions, and research organizations. In the many different uses cited, the most common outcomes are release/risk reduction, chemical substitution, increased communication, "good neighbor" agreements, laws/regulations, and improved emergency planning.

## A. USES OF EPCRA INFORMATION

**IBM Plant Agrees to Eliminate Use of CFCs - San Jose, CA - 1989:** An analysis of 1987 Toxic Release Inventory (TRI) data by the Citizens For a Better Environment showed that IBM's Silicon Valley plant was the largest emitter of chlorofluorocarbons (CFCs) in the state of California. The Silicon Valley Toxics Coalition organized local labor and environmental groups to pressure IBM for changes. In July 1989, a front page story in *USA Today* named IBM's Silicon Valley plant as the third largest (by volume) emitter of CFCs in the nation. Continuing to build support with newsletters, meetings, and contact with local enforcement agencies, the coalition proposed a "good neighbor" agreement that called for IBM to phase-out all use of CFCs. In September 1989, under the weight of increasing public pressure and negative publicity, IBM senior management announced a proposal to eliminate all use of CFCs in their products and processes by 1993. Not only did IBM switch to a safe substitute, but it asked their suppliers to do the same.[1]

**TRI Data Used to Compile a "Green Index" of Biggest Manufacturers - US - 1993:** Fortune Magazine compiled a "green index" of America's biggest manufacturers using TRI data as a central element. Fortune examined the environmental records of a number of companies, developing a relative ranking system that scored the companies from zero to 10 in 20 different performance categories, such as the amount of toxic emissions per dollar value of sales, and their percent reduction in toxic emissions.[2]

**Companies Switch to Less Dangerous Chemicals - Cuyahoga County, OH - 1990:** Using data made available through EPCRA, the Cuyahoga County Local Emergency Planning Committee (LEPC) conducted a hazard analysis of nearly 300 facilities that handle hazardous materials. A vulnerability zone was mapped for each facility, marking the area surrounding a facility that would be effected by a toxic chemical release. Each map was made available in the local public libraries. As a result of this heightened interest in the safety of surrounding neighborhoods, Cleveland's largest sewage treatment plant decided to eliminate a 55-ton railroad tank car of chlorine from its operations. The 1990 Annual Report of the Northeast Ohio Regional Sewer District announced the change from chlorine gas to liquid sodium hypochlorite, a safer disinfectant. One plant manager credited right-to-know with increasing awareness and a re-examination of chemical hazards that had been accepted as routine for years.[3]

**Good-Neighbor Agreement Successful - Berlin, NJ -** Using right-to-know data, the New Jersey Coalition Against Toxics asked five local facilities for the opportunity to inspect the plants for toxic hazards to workers and the community. Dynasil Corporation of America was the first to respond. An inspection team, made up of the local fire chief, members of the local emergency planning committee (LEPC), several neighbors, and two technical consultants toured the facility and made recommendations as to how Dynasil could improve worker safety and prevent a toxic disaster. The President of Dynasil made a company commitment to implement the LEPC suggestions and did so within a month. This cooperation represented one of the first good-neighbor agreements in the nation.[4]

**Pollution Prevention Through Worker-Management Agreements - New York -** The Citizens Environmental Coalition (CEC), a statewide citizens advocacy organization, uses TRI data for a number of citizen guides, fact sheets, and information packets. One of the most successful applications of TRI data by CEC involved a series of workshops that use these data to familiarize employees with hazards in the workplace. Many workers who attended the workshops are unaware of much of the TRI data and surprised at the emissions reported by the plants in which they work. However, workers and management have been able to open dialogues, even leading to emissions reductions. For example, Harrison Radiator in Lockport stopped using a number of hazardous chemical solvents because of pressure from workers. In addition, Kodak reduced emissions from 24 to 14 million pounds, in part as a result of the CEC workshops.[5]

**True Flexibility of TRI Data Shown Through Novel Uses - 1997:** An ever-increasing diversity of uses are being found for TRI data. Insurance companies, stock analysts, house hunters, epidemiologists, journalists, and all those rating America's best cities are finding the TRI data valuable. One can now draw correlations, for example, between cancer rates and the amount of carcinogen releases in a state or local community. In addition, the Detroit News was one of the first organizations to use the TRI data to examine the "environmental justice" debate, suggesting that big polluters tend to be located in low-income communities.[6]

**Right-to-Know Empowers Citizens - Contra Costa County, CA - 1989:** In February 1989, Communities for a Better Environment (CBE), a California non-profit, used right-to-know data to develop a report highlighting the threat of a toxic chemical incident in Contra Costa County. The report identified sixty-five companies in the area that collectively stored 140 million pounds of highly hazardous chemicals. CBE created and distributed a leaflet that summarized TRI data and described the location of the storage areas and amount of each chemical stored throughout the county. CBE also criticized the County Health Department for failing to ask local facilities for Risk Management Prevention Plans (RMPP), as required by state law.

Following an explosion and fire at a Chevron facility in April 1989, the CBE and the West County Toxics Coalition wrote to the County Health Department requesting that RMPPs be required from chemical companies. Armed with data from their right-to-know analysis, the Coalition and CBE appeared before the County Board of Supervisors and demanded the RMPPs be required for public safety. The Board agreed, and the first RMPP requests went out on December 1, 1989.[7]

**EPCRA Leads to Decreasing Chemical Hazards in Florida Communities - Florida - 1999:** EPCRA has led to a number of changes in Florida communities that have lowered the chemical-related risks to local citizens. The requirement to perform a hazard analysis under EPCRA section 302 led several facilities to work more closely with the local fire department to minimize or eliminate the risk of a spill. A number of water treatment facilities in the Melrose/Keystone Heights area switched from chlorine to a hypochlorite solution, especially important in smaller communities with limited hazmat response capabilities. Many swimming pools in cities such as Gainesville now limit the total chlorine that can be stored on-site at any given time. Facility reviews by the local emergency planning committees (LEPCs) have also spurred the elimination of pressurized regulators on chlorine systems and a change towards safer vacuum systems, which are now standard. LEPCs, such as the North Central Florida LEPC, are now reviewing emergency response information provided in facility risk management plans, which will supplement site-specific information already included in the LEPC emergency plan.[8]

**TRI Catalyst for Emission Reductions - North Carolina - 1990:** This article claims that the release of TRI data to the public was the catalyst that led to passage of the first air toxics regulations in North Carolina. After the NC Environmental Defense Fund announced that companies had legally released over 100 million pounds of toxic substances in 1987, Governor James G. Martin backed the regulations, which took effect in May 1990. These control-oriented standards, set by the Environmental Management Commission, required hundreds of industries to reduce emissions of 105 toxic air pollutants.[9]

**Business Community Begins to See Economic Benefits Related to TRI - 1996:** Since 1988, national environmental groups have been using TRI data to identify the top polluters in the United States. Economic incentives pushed major corporations to quickly reduce emissions to avoid developing a reputation as a major polluter. Armed with TRI data, local citizen groups are now able to document their concerns and force companies to address the risk of chemical spills.

Despite initial reservations, many business sectors now view the disclosure of TRI and EPCRA information as important for public outreach and monitoring of performance. Companies that previously may have been unaware of the extent of their releases are motivated to reduce emissions that can now be accurately tracked by citizens groups.

Since TRI's beginnings, an increasing number of businesses are recognizing the positive link between public accountability and business performance. In recent years, more than 100 companies have begun issuing annual environmental reports to investors, communities, environmental groups, and government. The reports, based primarily on TRI data, describe environmental goals, achievements, and setbacks.[10]

**NICs *Scorecards*™ Help Explain TRI Data - Charleston, WV - 1994:** The National Institute for Chemical Studies (NICs) has summarized TRI data in a more accessible format to help improve the emergency preparedness efforts of state emergency response commissions (SERCs) and local emergency planning committees (LEPCs). Because TRI data can help communities better understand and manage environmental risks, NICs developed *Scorecards*™, customizable for each locale, that offer the SERCs and LEPCs an excellent avenue for the public dissemination of toxic release information. The scorecards compress and interpret TRI data and allow companies to explain the details behind those numbers, facilitating better communication and understanding.[11]

**Labor Union and Community Groups Unite, Company to Reduce Emissions and Use - Northfield, MN - 1990:** The naming of Sheldahl Inc. as one of the nations leading emitters of airborne carcinogens coincided with contract negotiations between Sheldahl and the Amalgamated Clothing and Textile Workers Union (ACTWU). The announcement also led to the formation of two new citizens groups advocating pollution reduction and public health. The union had long been trying to reduce worker exposure to methylene chloride, and the increased concern for public safety added new weight to the negotiations. The union included environmental issues in their new contract negotiations. In an effort to ensure that public concerns did not shut the plant down, the union insisted that local citizens groups be present for the pollution negotiations. The resulting agreement phased out the use of methylene chloride in production by switching to a non-toxic substitute and called for a 90 percent emissions reduction by 1993.[12]

**TRI Triggers Emissions Monitoring Network and Reductions - Rochester, NY - 1992:** Spurred by 1989 TRI data indicating that its Rochester, NY facility ranked second in the United States for emissions of dichloromethane (DCM), Eastman Kodak pledged to cut DCM emissions by 70 percent by 1995. To track its progress in meeting these goals, Kodak stepped up emissions gauging by implementing an air emissions monitoring network.

Timing the startup of the monitoring program with a plant expansion and a company desire to focus on community safeguards, the community was kept well informed during public meetings held during permit review. Kodak also planned to analyze methanol, acetone, ethanol, and toluene releases.[13]

**Web Site Provides Data for Pollution Prevention - Elyria, OH - 1998:** After suffering from poor health that appeared to improve when she was away from home, Pauline Leboda of Elyria, Ohio began to suspect an environmental cause for her symptoms. She contacted Teresa Mills of the Buckeye Environmental Network. Mills used the Scorecard web site, sponsored by the

Environmental Defense Fund *(www.scorecard.org)*, which provides TRI pollution data based upon zip code, to discover what toxic emissions were causing the pungent odors in Elyria.

Further research proved that a local sponge manufacturer was operating without the proper toxic emissions permit. In a settlement with the EPA, the sponge manufacturer agreed to pay a fine and began using a chemical scrubber. Leboda and Mills proved that information can lead to action.[14]

**Largest Environmental Polluter in State Attempts to Reduce Waste - Derry, NH - 1997:**
After being identified as the largest environmental polluter in New Hampshire, senior management at HADCO Corporation initiated efforts to reduce releases of toxic chemicals and transfers of these substances from its facility. One of the nation's largest manufacturers of printed wiring boards, HADCO used chlorinated solvents in their multi-step manufacturing process. To reduce waste, HADCO implemented a solvent recovery system for some processes and eliminated the use of chlorinated solvents for others, while installing a continuous emissions monitoring system (CEMS). The company also switched to aqueous solvents, removing methylene chloride and other toxics from its process.

As a result of these actions, HADCO eliminated the annual disposal of 800,000 pounds of methylene chloride and no longer needed to operate the CEMS. Investment in this technology paid for itself in three years.[15]

**The Good Neighbor Project - Minnesota - 1991:** Following enactment of the 1990 Minnesota Toxic Pollution Prevention Act, over 500 Minnesota manufacturers were required to compose a pollution prevention plan and submit annual progress reports to the Minnesota Pollution Control Agency. The Good Neighbor Project, initiated in 1991 by the Minnesota Citizens for a Better Environment (CBE), encouraged community involvement in these pollution prevention plans. The Good Neighbor Project also identified the toxic polluters with the greatest potential environmental and health impacts and helped open a dialogue between the communities and industry.

In January 1993, CBE released "Get to Know Your Local Polluter," a report that profiled the top 40 polluters in Minnesota and provided communities with applicable TRI data, local demographics on nearby sensitive populations, exposure scenarios, and potential effects of the chemicals used or stored at each facility. The CBE is acting as the organizing body for many of these Good Neighbor agreements. By using right-to-know and TRI data, CEC is providing communities with the information they need to play a role in local industry pollution prevention and teaching industry how to work with the surrounding communities.[16]

**Wake-up Call Reduces Pollution and Saves Money - Emigsville, York County, PA - 1994:**
After Berg Electronics began reporting emissions under TRI, the company realized that they were releasing almost 300,000 pounds of hazardous chemicals into the environment. By installing a new cleaning system, the company reduced its emissions to 391 pounds per year. Although up-front costs for the new system were relatively high ($500,000), the company now saves about $1.2 million each year by avoiding cleanup and hazardous waste disposal costs.[17]

**TRI Data Used to Track Progress of the Big Three Auto Companies - Michigan - 1995:** The Ecology Center Toxics Reduction Project used TRI data to follow the pollution reduction progress of the "big three" auto companies. Ecology Center was able to use TRI data to prove flaws in emissions reduction analyses performed by these companies. In addition, the Center worked with the Great Lakes Auto Pollution Prevention Alliance to initiate discussions between plant management and local communities.

As a result of these discussions, the president of Auto Alliance International, which had been one of the largest polluters by volume in the state, committed to an aggressive solvent reduction program. The company's program will recapture solvents in the process, saving money and improving air quality in the long run.[18]

**Changing Users of TRI Data Reflect Benefits of EPCRA - 1990:** When the TRI data first became available in June 1988, industry topped the list of reviewers, primarily checking to see the accuracy of their own data and how their releases compared to competitors. EPA began to notice a change in the user trends at the close of 1989. More and more state agencies, environmental officials, health care institutions, and citizens were beginning to use the TRI data in their attempts to improve public health, local land use, and regulatory actions. The access to TRI data enabled citizens and environmental groups alike to push for risk reduction, pollution prevention, and stronger environmental laws.

For example, Louisiana used the TRI data to pass legislation aimed at cutting toxic air emissions in half by 1994. Many states, such as Illinois and Indiana, now require property sellers to disclose EPCRA-related information so that all buyers are aware of the past uses of the land. A South Texas school district even used TRI data to locate a safe site for a new elementary school. Around the country, citizens, environmental groups, and state and local authorities are working with these data to create an accurate picture of the chemical risks in their community.[19]

**TRI Yields Concrete Results, Researchers Find - U.S. - 1994:** As a result of making TRI data public and accessible, a majority of citizen groups and industry respondents surveyed in one study reported that *release reduction efforts were undertaken* at plants, and that meetings were prompted between industry and citizens. In addition, some facilities have signed "good neighbor agreements," which include release reduction goals and citizen monitoring rights. The researchers go on to state that "more voluntaristic approaches, built on forced leveling of the information playing field, are a supplement to regulation for problems such as accident prevention and toxics use reduction."[20]

      Due to space considerations, some other successful uses of EPCRA data are referenced below only briefly.

**Public Pressure Leads to Toxic Use Reduction Laws - Massachusetts and Oregon - 1990:** TRI data used by public interest organizations, citizens, and legislators.[21]

**Colorado Manufacturer Takes First Steps, Makes Good Neighbor Pledge - Boulder County, CO - 1991:** Community and company action based on high TRI ranking.[22]

**Citizens Win Funding for Refinery Monitoring - British Petroleum, Lima, OH - 1989:** The Facility was the biggest air polluter in the state, per TRI.[23]

**Regulations Spurred by TRI - Louisiana - 1990:** TRI data brought about public awareness, which forced state policymakers to work towards a significant reduction of air toxics.[24]

**TRI Data and GIS Used to Prioritize Pollution Prevention Effort - New Jersey - 1993:** New Jersey used TRI data with with geographic information systems (GIS) technology to map the data and impacts.[25]

**TRI Data Act as Catalyst, Address Environmental Injustice Issue - Ohio - 1995:** Environmental organization claims to show disproportionate waste and water impacts.[26]

**Manufacturer Cleans Up After Several TRI Appearances as a Top NY Polluter - Lockport, NY - 1996:** Action came after appearing in TRI reports as one of state's top polluters.[27]

**TRI Data Used to Change State Legislation - Utah - 1991:** TRI data helped Sierra Club and legislature identify problems.[28]

**EPCRA Data Help Detail Correct Response to Midwest Floods - 1993:** Local governments, emergency responders, companies benefitted from their contingency plans built around EPCRA data.[29]

**TRI Data Lend Support to "Environmental Justice" Debate - Chicago, IL - 1991:** Environmental group used TRI data to call attention to hazardous waste transporters from out of state.[30]

**Facilitating Cooperation Through Committee - Tennessee - 1994:** State that ranked high in TRI emissions formed state, industry, and non-profit committees to analyze TRI data, reduce pollution, and increase public awareness.[31]

**TRI Data Help Identify Health Risks - New York - 1995:** The State Department of Health used TRI data to develop rankings that suggested health risks that could result from toxic releases.[32]

**Publicity of TRI Proves To Be Motivating Factor - Westwego, LA - 1992:** American Cyanamid launched program aimed at cutting TRI emissions by 80 percent.[33]

**TRI Data Prove to be Powerful Tool in Stopping Polluters - Baton Rouge, LA - 1995:** An environmental organization uses TRI data to identify polluters and promote environmental justice.[34]

**Community/Manufacturer Agreement Improves Plant Safety and Relations - Manchester, TX - 1993**[35]

**TRI Data for U.S. Companies Used to Assess Facilities in Mexico - Texas - 1994**[36]

**Organized Community Defeats Ammonia Facility Application - Cloverleaf, TX**[37]

## B. USES OF RIGHT-TO-KNOW INFORMATION ASIDE FROM EPCRA

**States with Right-to-Know Programs Reduce Emissions Significantly - U.S. - 1995/1996:**
Several states have their own environmental right-to-know programs; studies of these can reveal impacts likely in national programs. In one study, researchers used data from the TRI to conduct a preliminary analysis of the effectiveness of right-to-know programs in reducing reported releases across the 50 states. They found that states with functional right-to-know programs are significantly more successful in reducing in-state toxic emissions that states without them. In addition, they found that the effect of a right-to-know program on toxic releases outside a state was not significant, contradicting an argument that such self-protection policies shift pollution to other states.[38]

**Toxics Use Down 20 Percent in Massachusetts - Boston, MA - 1997:** In the four years following enactment of the Toxics Use Reduction Act (TURA) of 1990 in Massachusetts, the use of toxic chemicals by companies in the state dropped 20 percent and the volume of chemicals ending up as waste also fell by 30 percent. While TURA does not mandate any process changes, it does require facilities to disclose what chemicals they use and the waste that they generate. The lack of any direct mandates for change in industry raised doubts among some companies and industry associations. However, the state government and environmental organizations such as MASSPIRG, argue that the public disclosure of TURA and TRI data seems to be the motivating factor for the reduction in toxic chemical use as manufacturers look for cost-effective and safer alternatives. Paul Burns of MASSPIRG noted that "the bottom line is that citizens have the right to know when and how they are exposed to toxic chemicals. This law [TURA] has been a powerful tool in getting businesses to voluntarily change their behavior."[39]

**Indonesia Plants Clean Up to Make Better Pollution Grade - Indonesia - 1999:** Indonesia's Environmental Impact Management Agency ran a pilot program, known as "PROPER," in which certain industrial facilities were graded, based on their water pollution performance. Researchers found that, "if [reputation effects] are important, then market agents and communities, *once properly and accurately informed*, can interact with firms to establish jointly-optimal levels of consumption and production."[40]

Disclosure of the Indonesian facilities' grades was sufficient to prompt 10 factories to invest in pollution abatement in order to improve their rating, and lead to a more than 40 percent pollution reduction in the pilot group in only 18 months. Other countries are adopting similar programs. In the Philippines, the national environmental agency's "EcoWatch" program has already used disclosure to dramatically increase compliance among 52 factories. It seems that if the U.S. were to significantly restrict right-to-know, it would run counter to a broad and growing international trend.[41]

**Impact of Public Information in Canada:** A 1999 study funded by the World Bank, *Incentives for Pollution Control: Regulation And (?) Or (?) Information*, analyzed regulatory enforcement, public information, and the relationship between the two (noted by the question marks in the title) regarding emissions in British Columbia. The study found that clear, strong standards with a significant and credible penalties produce emissions reductions. The authors also found that "the public disclosure of environmental performance does create *additional* and *strong* incentives for pollution control." They found evidence that the impact of public information was stronger than that of fines. They concluded that the combination of regulations and information puts different kinds of pressure on firms, "increasing the likelihood that they will undertake actions in line with environmental protection."[42]

**Appliance Labelling for Energy Consumption Encourages Environmental and Cost Savings - U.S. - 1995:** Ensuring that the public has access to information can be achieved in many ways. One way is product labeling. For instance, large appliances now carry labels describing expected energy usage and costs. It is well documented that, in the past, appliance buyers frequently did not buy energy-efficient devices, despite the fact that the rate of return due to energy cost savings far exceeded that of other investments. This phenomenon is attributed to multiple factors, one being the transaction cost of transferring information known by the seller to a potential buyer. Increasing the information available to consumers via standardized labels reduces this barrier, and can both benefit the environment and save the consumer money.[43]

**CONCLUSION**

Granted, labeling products differs from disseminating information about community hazards via paper or the Internet. However, the above examples show that improving the ease of acquiring such information clearly contributes to behavioral change by individuals and companies. While there are differences in the types of information and in the means of access in the programs above, the trend is clear. *Information that is readily available and relevant to the well-being of individuals often leads to improvements for the interested parties – and often for the larger public as well.*

# APPENDIX C REFERENCES

1. Tryens, Jeffrey, Richard Schrader, and Paul Orum. *Making the Difference: Using the Right-to-Know in the Fight Against Toxics*. Washington, DC: Center For Policy Alternatives and Working Group on Community Right-to-Know. (undated): p. 1. (Based on interview with Ted Smith, Silicon Valley Toxic Coalition.)

2. U.S. EPA. *An Overview of Uses of the Toxics Release Inventory Data in the U.S.*. Susan B. Hazen. Office of Pollution Prevention and Toxics. (June 1995): p.11. (Based on Faye Rice, "Who Scores Best on the Environment." *Fortune*, Vol.128, No.2. (July 26, 1993): p.114-122.)

3. Settina, Nina and Paul Orum. *Making the Difference, Part II: More Uses of Right-to-Know in the Fight Against Toxics*. Washington, DC: Center For Policy Alternatives and Working Group on Community Right-to-Know. (October 1991): p. 11-12.
Contact: Stuart Greenberg, Environmental Health Watch

4. Tryens, Jeffrey, Richard Schrader, and Paul Orum. *Making the Difference: Using the Right-to-Know in the Fight Against Toxics*. Washington, DC: Center For Policy Alternatives and Working Group on Community Right-to-Know. (undated): p. 15. (Based on article and interview with Jane Nogaki, New Jersey Coalition Against Toxics).

5. Michuda, Colleen. *TRI Success Stories: Citizens Making a Difference*. U.S. EPA. (August 1994): p. 6.
Contact: Diane Hemingway, Director, CEC - Western NY.

6. Selcraig, Bruce. "What You Don't Know Can Hurt You." *Sierra*. Vol. 82, No.1. (Jan-Feb 1997): p.38-45.

7.Tryens, Jeffrey, Richard Schrader, and Paul Orum. *Making the Difference: Using the Right-to-Know in the Fight Against Toxics*. Washington, DC: Center For Policy Alternatives and Working Group on Community Right-to-Know. (undated): p.14. (Based on interview with Michael Belliveau, Communities for a Better Environment).

8. Mundy, Dwayne. e-mail November 11, 1999.

9.Tryens, Jeffrey, Richard Schrader, and Paul Orum. *Making the Difference: Using the Right-to-Know in the Fight Against Toxics*. Washington, DC: Center For Policy Alternatives and Working Group on Community Right-to-Know. (undated): p.12-13. (Based on interview with Ed Norman, NC Environmental Defense Fund).

10. Hearne, Shelley. "Tracking Toxics: Chemical Use and the Public's Right-to-Know." *Environment*, Vol. 38, No. 6. (July-August 1996): p. 4-16.

11. *Right-to-Know Planning Guide*, "Emergency Planning..." Bureau of National Affairs, (May

1994): p. 2.

12. Settina, Nina and Paul Orum. *Making the Difference, Part II: More Uses of Right-to-Know in the Fight Against Toxics.* Washington, DC: Center For Policy Alternatives and Working Group on Community Right-to-Know. (October 1991): p. 3-4.
Contacts: Richard Metcalf, ACTWU, and Frank Wolf, Clean Air in Northfield.

13. *BNA Reporter.* "Preventing Pollution..." Bureau of National Affairs (February 27, 1992): p. 3.

14. "EDF Letter." Environmental Defense Fund. December 1999, Vol. 30, No. 6, p. 5.

15. *BNA Reporter.* "Preventing Pollution..." Bureau of National Affairs, (December 25, 1997): p. 3. (discussing an EPA case study)

16. Michuda, Colleen. *TRI Success Stories: Citizens Making a Difference*. U.S. EPA. (August 1994): p. 10-11.
Contact: Jo Haberman, Citizens for a Better Environment - Minnesota.

17. Michuda, Colleen. *TRI Success Stories: Citizens Making a Difference*. U.S. EPA. (August 1994): p. 17. (Based on article in "*Philadelphia Inquirer*," June 18, 1994.)

18. *The Right Stuff: Using the Toxics Release Inventory.* OMB Watch and Unison Institute. (July 1995): p. 17.

19. *Right-to-Know Planning Guide*, "EPA Notices Change in Users of Toxic Release Inventory." Bureau of National Affairs. (February 1990): p. 4.

20. Lynn, F. M. and J. Kartez. "Environmental Democracy in Action: The Toxics Release Inventory," *Environmental Management*, Vol. 18, No. 4, pp. 511-521, 1994, pp 517-519.

21. Tryens, Jeffrey, Richard Schrader, and Paul Orum. *Making the Difference: Using the Right-to-Know in the Fight Against Toxics.* Washington, DC: Center For Policy Alternatives and Working Group on Community Right-to-Know. (undated): p. 2. (Based on interview with Marc Osten, MASSPIRG and Quincy Sugarman, OSPIRG.)

22. Settina, Nina and Paul Orum. *Making the Difference, Part II: More Uses of Right-to-Know in the Fight Against Toxics.* Washington, DC: Center For Policy Alternatives and Working Group on Community Right-to-Know. (October 1991): p. 7-8.
Contact: Larry Bulling, Colorado Citizen Action.

23. Tryens, Jeffrey, Richard Schrader, and Paul Orum. *Making the Difference: Using the Right-to-Know in the Fight Against Toxics.* Washington, DC: Center For Policy Alternatives and Working Group on Community Right-to-Know. (undated): p. 9. (Based on interview with Ed Hopkins, Ohio Citizen Action).

24. Tryens, Jeffrey, Richard Schrader, and Paul Orum. *Making the Difference: Using the Right-to-Know in the Fight Against Toxics*. Washington, DC: Center For Policy Alternatives and Working Group on Community Right-to-Know. (undated): p. 12-13. (Based on interviews with Daryl Malek-Wiley, LA Sierra Club; Elouise Well, LA Environmental Action Network).

25. U.S. EPA. *An Overview of Uses of the Toxics Release Inventory Data in the U.S.*. Susan B. Hazen.Office of Pollution Prevention and Toxics. (June 1995): p.9.

26. *The Right Stuff: Using the Toxics Release Inventory*. OMB Watch and Unison Institute. (July 1995): p. 34.

27. *BNA Reporter*. "Preventing Pollution..." Bureau of National Affairs. (May 2, 1996): p. 3.

28. *The Right Stuff: Using the Toxics Release Inventory*. OMB Watch and Unison Institute. (July 1995): p. 28.

29. *Right-to-Know Planning Guide*. "EPCRA Data Plays Major Role in Midwest Flood Response." Bureau of National Affairs. (July 1993): p. 4.

30. *The Right Stuff: Using the Toxics Release Inventory*. OMB Watch and Unison Institute. (July 1995): p. 16.

31.Michuda, Colleen. *TRI Success Stories: Citizens Making a Difference*. U.S. EPA. (August 1994): p. 19. Contacts: Alan Jones, Tennessee Environmental Council and Angie Pitcock, Division Director, TN Dept. of Environment and Conservation.

32. U.S. EPA. *An Overview of Uses of the Toxics Release Inventory Data in the U.S.*. Susan B. Hazen. Office of Pollution Prevention and Toxics. (June 1995): p.7.

33. David Rotman. "Cleaning up processes," *Chemical Week*. Vol. 150, No. 23. (June 17, 1992).

34. *The Right Stuff: Using the Toxics Release Inventory*. OMB Watch and Unison Institute. (July 1995): p. 30-31.

35. *Right-to-Know Planning Guide*, "Partnership Linked to Improvements at Chemical Plant." Bureau of National Affairs. (August 1993): p. 4.

36. *The Right Stuff: Using the Toxics Release Inventory*. OMB Watch and Unison Institute. (July 1995): p. 19.

37. Settina, Nina and Paul Orum. *Making the Difference, Part II: More Uses of Right-to-Know in the Fight Against Toxics*. Washington, DC: Center For Policy Alternatives and Working Group on Community Right-to-Know. (October 1991): p. 15-16.

Contacts: Karla Land, North Channel Concerned Citizens Against Pollution; Jim Baldauf, Texans United; and Dr. Fred Millar, Friends of the Earth.

38. Grant, D.S. and Liam Downey. "Regulation though Information: An Empirical Analysis of the Effects of State-sponsored Right-to-know Programs on Industrial Toxic Pollution." *Policy Studies Review*, (Autumn/Winter 1995/1996): p 346.

39. *Right-to-Know Planning Guide*. "Massachusetts Reports 20 Percent Drop in Toxics Use." Bureau of National Affairs. (January 1997): p. 4.

40. Afsah, Shake, Benoit Laplante, and David Wheeler. "Regulation in the Information Age: Indonesian Public Information Program for Environmental Management." (March 1997): p. 5, 8. http://www.worldbank.org/nipr/work_paper/govern/index.htm.

41. World Bank, "Greening Industry: New Roles for Communities, Markets, and Governments." Oxford University Press. (1999): pp 64-74.

42. Foulon, J., P. Lanoie, and B. Laplante. "Incentives for Pollution Control: Regulation And (?) Or (?) Information," (October 1999) http://www.worldbank.org/nipr/work_paper/andor/index.htm.

43. Howarth and Sanstad, "Discount rates and energy efficiency," *Contemporary Economic Policy*. Vol. 13, No. 3. (July 1995): p. 101(9).

**APPENDIX D**

**HOW DOES NEGATIVE PRESS AFFECT
FACILITIES' TOXIC RELEASE EMISSIONS?**

# APPENDIX D

# HOW DOES NEGATIVE PRESS AFFECT FACILITIES' TOXIC RELEASE EMISSIONS: AN INFORMAL ANALYSIS

## Introduction

The Toxic Release Inventory (TRI) database, which requires manufacturing facilities that release to the environment any of 300 chemicals to file an emissions report with EPA, is a major component of the Emergency Planning and Community Right-to-Know Act (EPCRA). Every citizen can access this database and can determine the presence and release of hazardous and toxic chemicals at industrial locations. In this analysis, we wanted to discover whether and to what extent negative press coverage might affect a facility's TRI emissions. Does publicly provided information on toxic chemical releases lead to emissions reductions? When newspapers target a facility for its large amount of toxic emissions, does that facility reduce its toxic releases in subsequent years? Does newspaper and media criticism change corporate behavior and the allocation of corporate resources to lessen the negative criticism? This Appendix describes the results of an informal, case-based analysis on whether media criticism appears to have any positive effect on toxic emissions. It is not intended to represent a robust statistical analysis of this hypothesis.

## Approach

Several different searches for newspaper, magazine, or trade journal articles that cited Toxic Release Inventory data on one or more specific facilities were performed. The search criteria included keywords such as routine or annual toxic emission, toxic releases, TRI or Toxic Release Inventory or Section 313 or EPCRA or Community Right-To-Know, etc. Although a number of facilities were mentioned in various articles, seven facilities seem to have been cited most often in article after article, year after year, as "the worst polluters" in the country, according to their TRI emissions. In addition, a number of facilities were identified that were cited as "the worst polluters" in their states, according to their TRI emissions. The years in which these facilities were targeted by the media for their high level of emissions were also noted. The newspaper articles identified these facilities for the sheer quantity of their TRI emissions. For example, one facility could be responsible for 80 percent of a state's total TRI emissions in a one-year period.

The TRI database was used to compare the quantity of toxic release emissions for each of the seven "worst polluters" before it received negative press to the quantity of toxic release emissions after it received negative press. Obviously, there are many reasons for a company to reduce its emissions. Companies with large emissions may be able to find ways to reduce those emissions more quickly and easily than others and sometimes facility modifications to reduce emissions could take more than a year; some of these emission reductions may have been in the planning stages prior to negative press coverage. Consequently, emission results for several years after press coverage were included to see whether reductions in TRI emissions occurred sometime after a facility received negative press.

To further judge the effect of negative press, we identified a comparable facility located in the same region with the same Standard Industry Classification (SIC) code emitting the same chemicals as each of the "worst polluter" facilities in particular states. However, these comparable facilities did not receive negative press for their TRI emissions. For these comparable facilities, we used the TRI database to track the quantity of toxic release emissions over the same time period as the "worst polluting" facilities. We then compared the direction and rate of change in TRI emissions for publicized facilities as compared to non-publicized facilities. As above, emission reduction efforts can take time and can become more challenging as emissions are reduced; a facility may have been planning emission reductions regardless of press coverage.

## Results

### Nationwide

A significant number of companies have made a wide variety of reductions in their TRI emissions. However, as noted above, seven companies were selected for closer examination because they repeatedly appeared in articles year after year because of the large quantities of toxic chemicals they emit. As early as 1989, Du Pont, Monsanto, American Cyanamid, Kennecott, and IMC-Agrico were targeted as the worst polluters in the country based on their total emissions in 1987. These companies appeared in newspapers all over the country, from the *Los Angeles Times*, to the *St. Louis Post Dispatch*, to the *Lexington-Herald Leader*. A public interest group, Citizen Action, named Inland Steel a top ten toxic polluter in 1991 and the *Minneapolis-St. Paul Business Journal* named 3M among the nation's top polluters in 1991.

For a core group of chemicals, TRI emissions for nearly all companies have gone down 43% between 1988 and 1997. The core group of chemicals have remained the same since the first TRI reports in 1987. In contrast, the companies which were publicly criticized for large quantities of TRI emissions significantly reduced their emissions after receiving negative press. Those companies named as some of the "worst polluters in the country" reduced their emissions as much as twice the general TRI trend.

Between 1990 and 1996, the following companies significantly reduced their total toxic releases and transfers. These reductions are much greater than the general trend in TRI emissions reductions between 1991 and 1996.

## Table D-1 – Reductions of TRI Emissions for Selected Companies

| Company | Percent Reduction | General Trend (%) | Reduction over General Trend (%) | Increased Improvement over General Trend (factor) |
|---|---|---|---|---|
| Inland Steel | 95 | 43 | 52 | 2.2 times |
| Kennecott | 90 | 43 | 47 | 2.1 |
| Monsanto | 84 | 43 | 41 | 2.0 |
| American Cyanamid | 83 | 43 | 40 | 1.9 |
| IMC-Agrico | 82 | 43 | 39 | 1.9 |
| Du Pont | 73 | 43 | 30 | 1.7 |
| 3M | 65 | 43 | 22 | 1.5 |

Although a facility may reduce its toxic emissions for a number of reasons, "manufacturers listed among the worst polluters ... may change their ways out of fear of customer boycotts, increased regulation, or community hostility. The company's reputation, hard to build and easy to destroy, is at stake" (D55).

Ohio

Between 1989 and 1992, Ohio newspapers named Honda of America and O.M. Scott and Sons the highest polluters in the state. In 1989, the *Columbus Dispatch* reported that O.M. Scott was top polluter because of its ammonia releases. In 1991, the *Cleveland Plain Dealer* reported on Honda's toxic releases, the highest in the state. The *Columbus Business First* criticized both Honda and O.M. Scott for their toxic releases in a 1992 article. In addition, Ohio EPA cited Honda of America for nine air pollution violations in 1992. Between 1991 and 1994, Honda reduced its total toxic releases and transfers by 30 percent. However, since 1994, total on-site emissions by Honda have continued to decrease although off-site transfers have increased beyond the 1991 levels. O.M. Scott and Sons (now Scotts Company) reduced its total toxic releases and transfers 52 percent between 1990 and 1994. O.M. Scott's emission levels rose again in 1995 to almost the 1991 level, but declined again in 1996 and 1997.

Tennessee

In 1992, *USA Today* named the top 15 "toxic offenders" that release the most toxic material (in pounds). Four of Du Pont's facilities [New Johnsonville, Tennessee (6th of the 15), Beaumont, Texas (11th), Pass Christian, Mississippi (12th), and Victoria, Texas (14th)] were included in the list. Long named the nation's top polluter, Du Pont is cited in the press more than any other company for its toxic emissions. However, between 1991 and 1996, Du Pont reduced its emissions by 73% company-wide.

In 1991, the *Memphis Commercial Appeal* reported on Du Pont's pollution-control programs. In 1992, the *Commercial Appeal* reported that Tennessee was the nation's third largest producer of toxic waste in 1990, with Du Pont among the state's top 10 dischargers. But, in 1994, the *Memphis Business Journal* openly criticized Du Pont's New Johnsonville plant for the largest emissions of toxic chemicals in the state. Du Pont's New Johnsonville plant reduced its total toxic releases and transfers by 97 percent between 1990 and 1996; 94 percent of this was between 1994 and 1995. In 1991, the New Johnsonville plant accounted for 19% of Tennessee's total toxic releases. In 1996 however, this same plant accounted for 1 percent of Tennessee's total toxic releases. Certainly, negative press coverage appears to have helped spur corporate spending and improved behavior. Note that a company may need time to enact modifications between when negative press is published and changes in the amount of emissions.

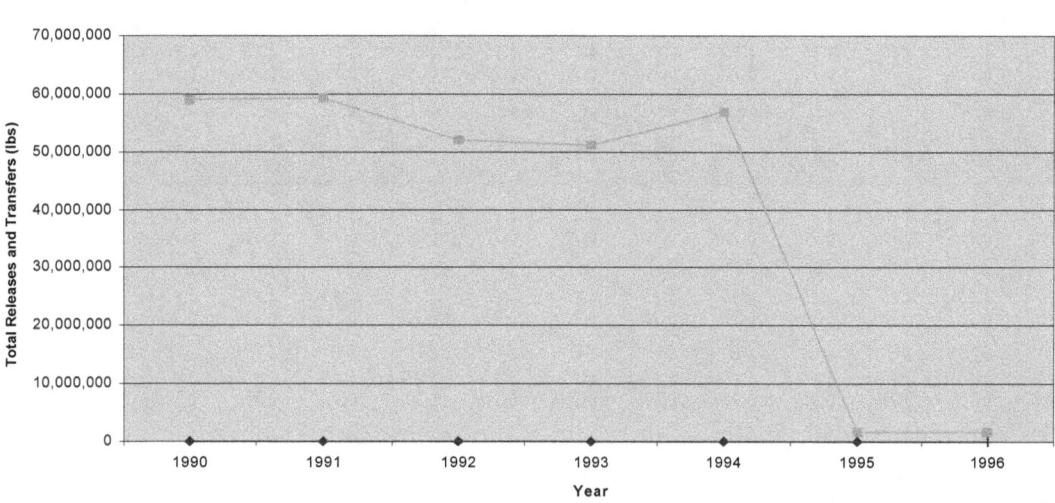

Du Pont, New Johnsonville, TN

We compared Du Pont's New Johnsonville plant to Kemira Pigments, Inc. in Savannah, Georgia. Both plants have the same SIC code (2816, Inorganic Pigments), they emit the same chemicals, and they are in the same EPA region. Between 1991 and 1996, Kemira Pigments emissions did not change significantly. Kemira Pigments was neither cited nor criticized in the press as was Du Pont.

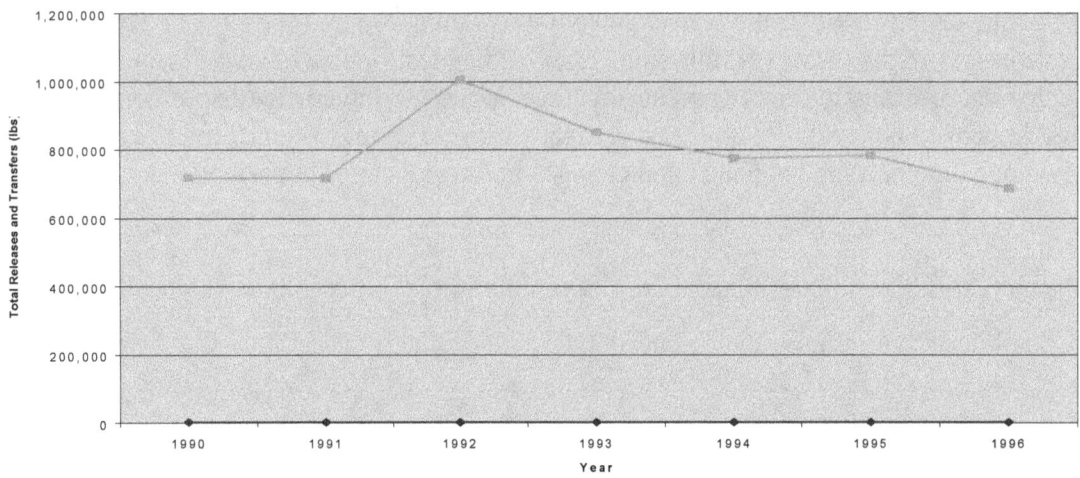

**Kemira Pigments, Inc., Savannah, GA**
**SIC 2816 (Du Pont, TN)**

## Montana

Asarco Inc.'s East Helena plant has been publicized as the top emitter of toxic releases in Montana since 1990. *The American Metal Market* reported in September 1992, that Asarco "ranked seventh on the list of manufacturing companies with the most toxic chemical releases." The company defended its ranking by explaining that "eighty-seven percent of its releases are made up of slag, which when properly managed presents no hazard to the public." *The Billings Gazette* reported that Asarco's East Helena plant led the 1993 list of Montana companies in discharges into the air, water and land in the course of manufacturing. Between 1990 and 1996, the East Helena plant <u>increased</u> its toxic emissions by 12 percent, while company-wide, Asarco Inc. <u>increased</u> its total toxic emissions by 50%. On average, Asarco, Inc. East Helena accounts for 92 percent of Montana's total toxic emissions between 1990 and 1996.

The Exxon Billings Refinery has also been in the Montana press for its toxic emissions. While *The Billings Gazette* named Asarco for the quantity of toxic chemicals it released, Exxon was named in 1991 for its release of xylene. Xylene can cause birth defects, and at high levels can cause dizziness, passing out and even death. Between 1990 and 1996, Exxon reduced its total xylene releases and transfers by 86 percent. Over the same time period, the Exxon Billings Refinery reduced its total releases and transfers by 64 percent. Whereas, for all of Exxon's facilities, the total toxic releases and transfers has stayed the same between 1990 and 1996. Company-wide, Exxon has not improved their total toxic emissions. But, in Montana, where Exxon received negative press, the Billings Refinery reduced their total TRI emissions.

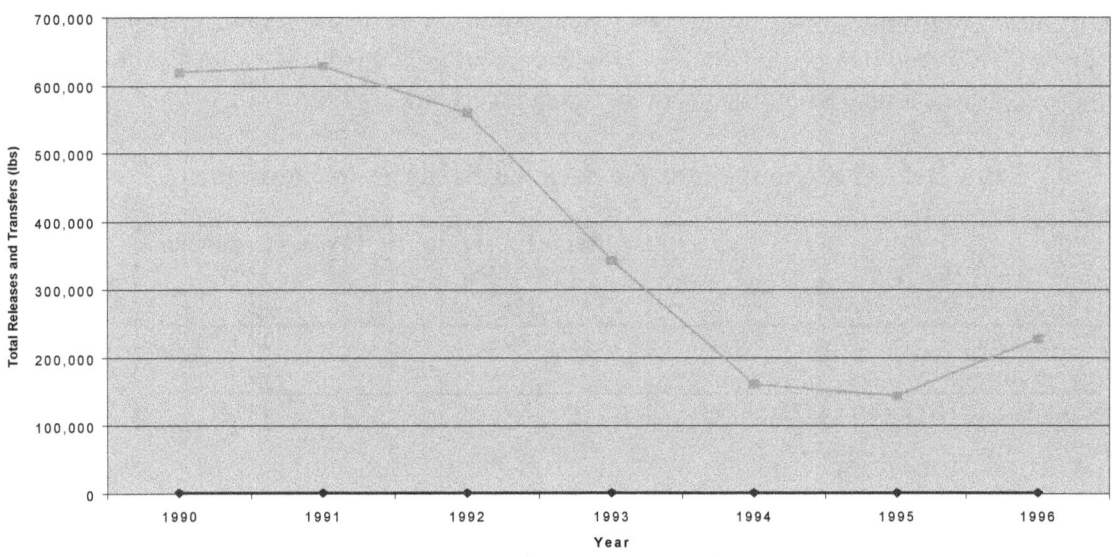

Exxon Billings Refinery, Billings, MT

## Louisiana

As early as 1987, Louisiana newspapers named a portion of the state "Cancer Alley." 15 chemical companies reside along a stretch of the Mississippi River near St. Gabriel. Residents blamed the pollution from these facilities for the area's "alleged high miscarriage, stillbirth, and cancer rates" according to a 1987 article in *Industry Week*. Since the first TRI data became available to the public in 1989, Louisiana has had the highest toxic emissions in the country. Four facilities in "Cancer Alley," American Cyanamid Fortier Plant, Agrico Chemical Faustina, Agrico Chemical Uncle Sam, and Arcadian Fertilizer, account for approximately 60 percent of Louisiana's total toxic releases. Not only was Louisiana named the worst polluting state in the country, but these four facilities were frequently in the Louisiana newspapers between 1989 and 1994.

In 1990, The *New Orleans Times Picayune* had two articles on the top polluters in Louisiana. One article states, "American Cyanamid's [facility in Waggaman] releases are greater than those of the entire state of New Jersey. Since New Jersey is 15th (in the nation) in total emission, this means American Cyanamid's releases are greater than the total releases of most states." In 1991, four articles in the *New Orleans Times Picayune* reported on Agrico's and Arcadian's high toxic emissions. Also in 1991, the *Picayune* called Louisiana "the Wasteland." They reported that American Cyanamid "discharged more toxic waste underground than the total discharged by any one of the 48 states in 1988." In 1992, the Picayune again criticized American Cyanamid for its "dubious distinction of being the most polluting plant in the country." Finally, in 1993 and 1994, the *Louisiana Industry Environmental Alert* reported on the total TRI emissions of Agrico, Arcadian, and American Cyanamid.

The total toxic emissions for Agrico Faustina and Agrico Uncle Sam steadily increased between 1990 and 1993. Arcadian Fertilizer increased its toxic emissions between 1990 and 1992. All of these companies drastically reduced their emissions by 1994. Between 1990 and 1994, American Cyanamid reduced its total toxic releases and transfers by 87%. All four of these facilities have kept their total TRI emissions at the same levels since 1994.

### Table D-2 – TRI Reductions for Selected Facilities in Louisiana

| Facility | Percent Increase (pre-1994) | Percent Reduction (by 1994) |
|---|---|---|
| Agrico Faustina | 126 | 90 |
| Agrico Uncle Sam | 71 | 94 |
| Arcadian Fertilizer | 97 | 68 |
| **LOUISIANA, Total Releases and Transfers** | **56** | **67** |

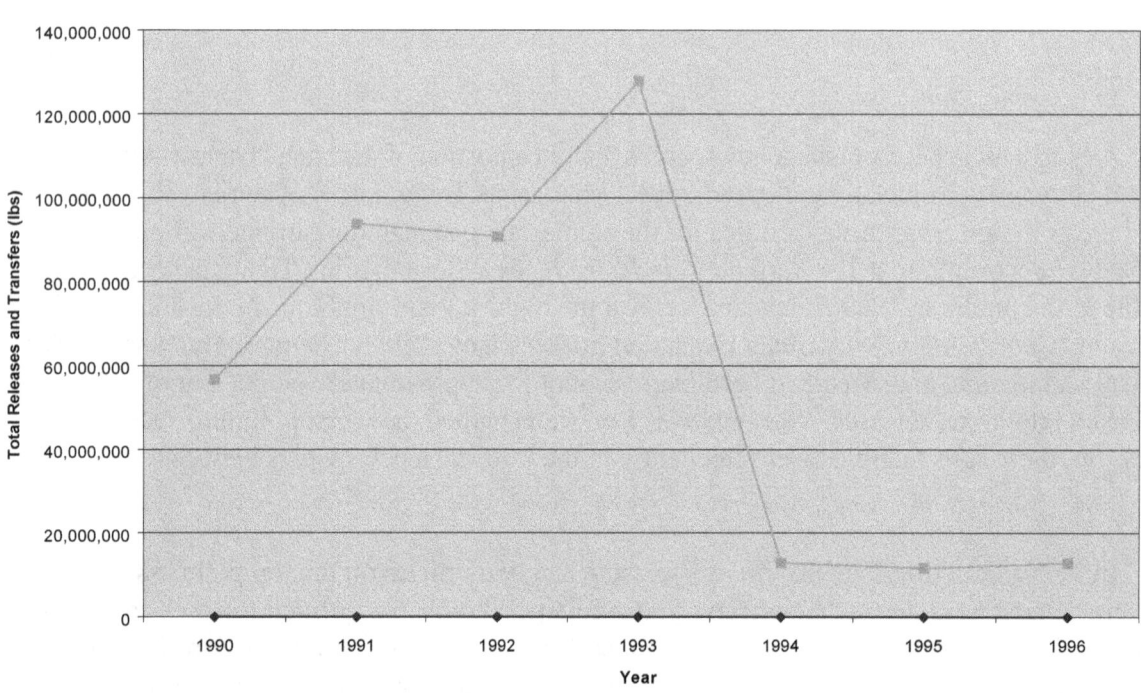

IMC-Agrico Faustina Plant, Saint James, Louisiana

**IMC-Agrico, Uncle Sam, Lousiana**

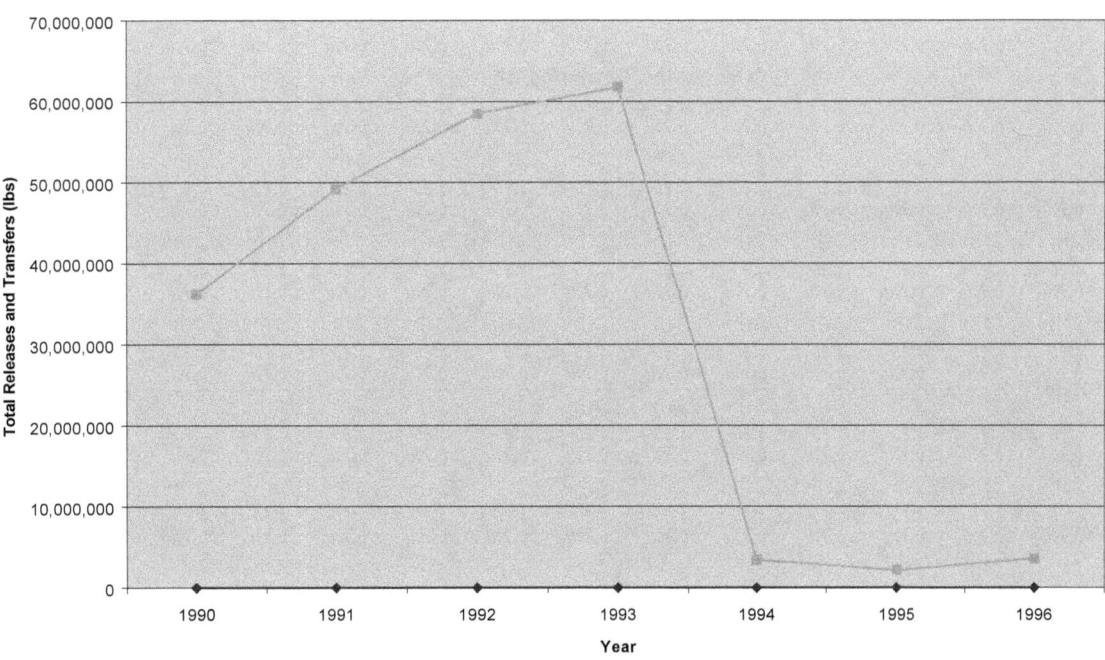

**Arcadian Fertilizer, Louisiana**

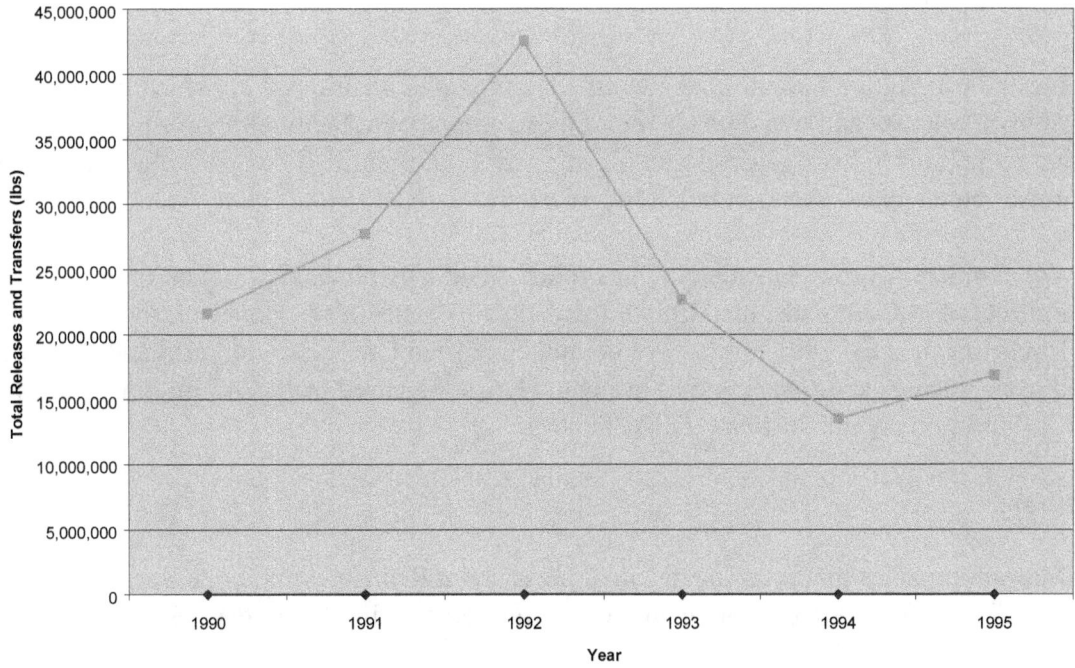

D-9

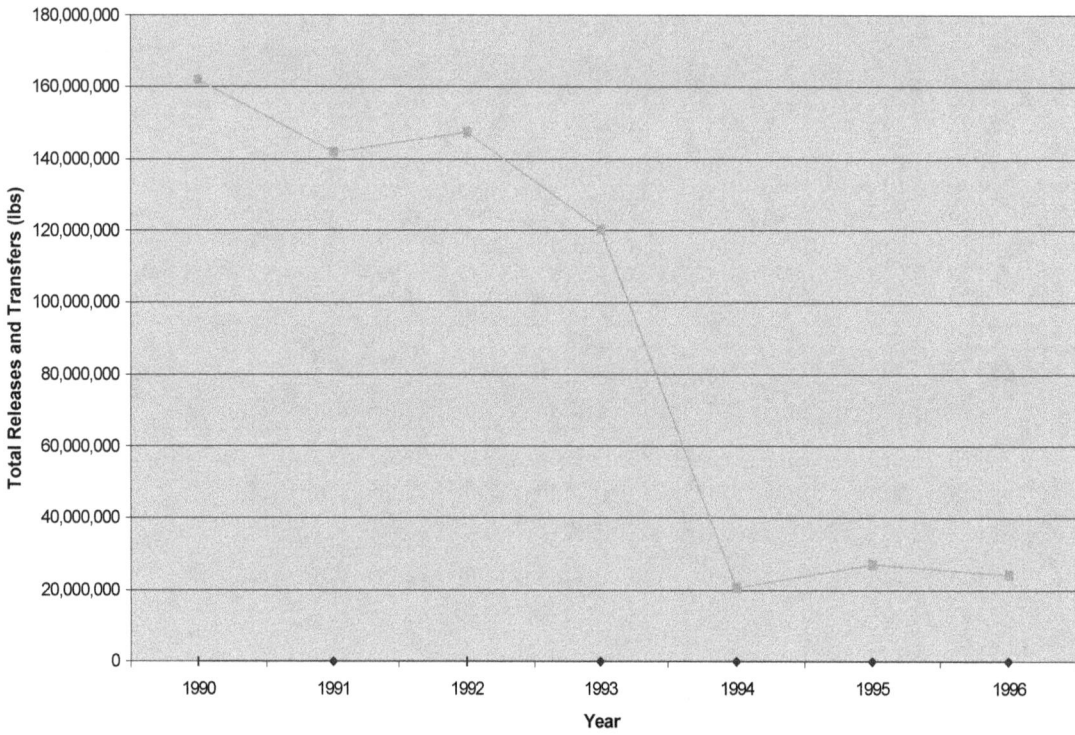

**Cytec Ind. Inc., Westwego, LA**
**(American Cyanamid)**

For the two Agrico facilities, we identified comparable facilities with which to compare their total toxic releases and transfers. None of these comparable facilities showed such significant reductions in toxic releases and transfers as those identified "worst polluters." Also, these comparable facilities were not publicly criticized.

We compared Agrico Faustina to Air Products and Chemicals, Inc. Total toxic releases for Air Products and Chemicals, Inc. declined slightly 1991 and 1994. However, their 1995 toxic release levels matched the 1991 levels. We compared Agrico Uncle Sam to Farmland Hydro L.P. The total toxic releases and transfers for Farmland Hydro increased from 1.47 million pounds in 1991 to 2.37 million pounds in 1996, a 61% increase.

**Conclusions**

- Many companies that received negative press about their total toxic releases reduced their emissions 1.5 to 2 times more than the general TRI trend in toxic releases.

- As seen above, in Tennessee, Du Pont dramatically reduced its total toxic releases after receiving negative press. Company-wide, Du Pont has also reduced its total toxic emissions.  The total releases of a facility comparable to Du Pont's New Johnsonville plant (but which did not receive negative press), generally did not change.

- In Montana, Asarco's East Helena plant increased their emissions after they received negative press.  However, the emissions increases at the Montana plant were slight (12 percent) compared to Asarco's company-wide total emissions (50 percent).  The Exxon Billings Refinery was named in the press for the quantity of xylene it released into the environment.  After this publicity, the refinery reduced the release of xylene by 86 percent and overall emissions by 64 percent.

- In Louisiana, the top four polluting facilities were frequently cited by the press between 1989 and 1993.  By 1994, all of these facilities had significantly reduced their total toxic releases.  Also, when we compared facilities, those that did not receive negative publicity did not significantly reduce their total toxic releases.

- Nearly all individual facilities studied reduced their TRI emissions well beyond the national trend in years following negative press coverage about their toxic emissions.  In some instances, negative press coverage regarding many facilities within a company preceded significant reductions in that company's overall TRI emissions.

# APPENDIX D REFERENCES

D1. Abrahamson, Peggy. "Group Cites PD for Pollution." *American Metal Market*. The Gale Group. DIALOG File 148. July 17, 1991.

D2. "America's Worst Toxic Polluters; 8 Companies with Poor Environmental Records." *Business and Society Review*. The Gale Group. DIALOG File 148. Winter, 1993.

D3. Anderson, Ed. "Plant's Emissions Outrank Entire States." *New Orleans Times Picayune*. DIALOG File 706. December 11, 1990.

D4. _____. "Six Area Plants Among Louisiana's Top Polluters." *New Orleans Times Picayune*. DIALOG File 706. April 28, 1990.

D5. Associated Press. "Group Releases 'Toxic 500' List: 24 Plants Produce a Third of the Hazardous Waste, Statistics Show." *The Orlando Sentinel*. DIALOG File 705. August 11, 1989.

D6. Bertelson, Christine and Ahmed, Safir. "Clean Air: States Vary Approach." *St. Louis Post Dispatch*. DIALOG File 494. January 21, 1990.

D7. Brookes, Warren. "Pollution Curb, Learning Curve." *The Washington Times*. DIALOG File 717. January 9, 1991.

D8. Brown, T.C. and Tatge, Mark. "Honda's Home No. 1 in Ohio for Toxic Air." *The Cleveland Plain Dealer*. DIALOG File 725. December 19, 1991.

D9. Bukro, Casey. "Council Criticizes 8 Firms' Records on Environment." *The Chicago Tribune*. DIALOG File 632. December 15, 1992.

D10. Charlier, Tom. "Big Polluters in Shelby Cutting Back Toxic Flow." *The Memphis Commercial Appeal*. DIALOG File 740. July 22, 1991.

D11. _____. "Tennessee is Third in Toxic Waste." *The Memphis Commercial Appeal*. DIALOG File 740. May 28, 1992.

D12. Dias, Monica. "Report Cites Area Polluters State's Worst Include 2 Local Companies." *The Kentucky Post*. DIALOG File 722. July 23, 1991.

D13. Downing, Bob. "Report Details Release of Toxic Chemicals in Ohio for Year." *Akron Beacon Journal*. Resp. Database Services. DIALOG File 9. March 26, 1995.

D14. "Du Pont Reports 6% Rise in Toxic Chemical Releases." *The Wall Street Journal*. May 18, 1993.

D15. "Du Pont Reports Decline in 1990 Toxics Releases." *Chemical Marketing Reporter*. The Gale Group. DIALOG File 16. July 8, 1991.

D16. Edwards, Randall. "Union County Gets Bad Rap." *The Columbus Dispatch*. DIALOG File 495. June 14, 1989.

D17. Epstein, Robin. "Ranking the Heavy Emitters: Toxic Corporations." *The Nation*. The Gale Group. DIALOG File 88. December 5, 1994.

D18. "FOE Report Claims Du Pont Worst Polluter." *Chemical Marketing Reporter*. The Gale Group. DIALOG File 16. September 2, 1991.

D19. Fried, John J. "Who's Who in Toxic Emissions." *The Philadelphia Inquirer*. DIALOG File 633. August 28, 1992.

D20. "Industrial Wastes: State-By-State Look." *USA Today*. DIALOG File 703. May 28, 1992.

D21. Johnson, Clair. "Ranking of Polluting Firms Worsens." *The Billings Gazette*. Bell & Howell. DIALOG File 635. No date.

D22. Journal of Commerce. "A Few Factories Release Most of Toxic Pollutants, Study Says." *Sacramento Bee*. DIALOG File 496. October 6, 1990.

D23. _____. "Bulk of Toxic Pollution Tied to Handful of Firms: EPA Finds 36% of Emissions Spewed by 50 Plants." *Lexington Herald-Leader*. DIALOG File 721. October 6, 1990.

D24. Kurschner, Dale. "3M's Pollution Record: Cloudy and Clear." *Minneapolis-St. Paul City Business*. The Gale Group. DIALOG File 148. December 9, 1991.

D25. Lambrecht, Bill. "Polluters Ranked." *St. Louis Post Dispatch*. DIALOG File 494. July 12, 1991.

D26. Lombardi, Bill. "39.1 Million Pounds of Toxic Chemicals Released." *The Billings Gazette*. Bell & Howell. DIALOG File 635. No date.

D27. Los Angeles Times. "Toxic Chemical Releases Down by 9%." *St. Paul Pioneer Press Dispatch*. DIALOG File 701. October 4, 1990.

D28. _____. "Toxic Waste Discharges are Reduced." *St. Paul Pioneer Press Dispatch*. DIALOG File 701. October 4, 1990.

D29.  Lucas, Allison.  "Du Pont:  Fighting its Way to a Cleaner Standard:  But Pollution Numbers are Slow to Come Down."  *Chemical Week*.  The Gale Group.  DIALOG File 148.  June 1, 1994.

D30.  Martz, Michael.  "Prevention Seen as Profitable:  Du Pont Finds Resource in Recycled Solvent."  *The Richmond News Leader*.  Richmond Times Dispatch.  DIALOG File 709.  July 26, 1990.

D31.  "Metals Rated for Pollution and for Improvement."  *American Metal Market*.  The Gale Group.  DIALOG File 16.  April 26, 1994.

D32.  "Monsanto, Olin and Du Pont Claim Sharp Emissions Cuts."  *Chemical Marketing Reporter*.  The Gale Group.  DIALOG File 16.  July 15, 1991.

D33.  Moos, Shawna.  "Pollution-Prevention Power to the People."  *Technology Review*.  Bell & Howell.  DIALOG File 15.  October, 1992.

D34.  Nelson-Horchler, Joani.  "Superfund Poses Major Problems for Industry."  *Industry Week*.  The Gale Group.  DIALOG File 148.  September 21, 1987.

D35.  O'Byrne, James and Schleifstein, Mark.  "In Harm's Way – Geismar/St. Gabriel By the Numbers."  *New Orleans Times Picayune*.  DIALOG File 706.  February 18, 1991.

D36.  _____.  "Louisiana Industries Nearly Halved Pollution, DEQ Report Says."  *New Orleans Times Picayune*.  DIALOG File 706.  October 4, 1991.

D37.  _____.  "Louisiana Leads U.S. In Toxins."  *New Orleans Times Picayune*.  DIALOG File 706.  May 28, 1992.

D38  _____.  "Louisiana's Top Ten Toxic Polluters*."  New Orleans Times Picayune*.  DIALOG File 706.  February 18, 1991.

D39.  _____.  "Troubled Plant:  Neighbors Would Like to Breathe Easier."  *New Orleans Times Picayune*.  DIALOG File 706.  February 18, 1991.

D40.  _____.  "Underground Hazards:  Drinking Water Fears Spread With Wastes."  *New Orleans Times Picayune*.  DIALOG File 706.  March 25, 1991.

D41.  Phillips, Michael M.  "The Price of Industry:  Tons of Pollutants 10 Million Pounds Released in 1987, EPA Records Show."  *The Philadelphia Inquirer*.  DIALOG File 633.  April 26, 1989.

D42.  Pope, Charles.  "State Moves Up the List as Polluter:  EPA Ranks State 23rd in Toxic Emissions."  *The State*.  DIALOG File 720.  May 17, 1991.

D43. "Responsible Care: American Cyanamid." *Chemical Week*. The Gale Group. DIALOG File 16. June 17, 1992.

D44. Rice, Faye. "Who Scores Best on the Environment." *Fortune*. Bell & Howell. DIALOG File 15. July 26, 1993.

D45. Roberts, Timothy. "Tennessee Firms Work to Shed State's 'Dirty' Image." *Memphis Business Journal*. The Gale Group. DIALOG File 148. July 11, 1994.

D46. Rutherford, Glenn. "Kentucky 21st in Pollution , Group Says Jefferson, Marshall, Daviess Counties Emis Most Toxins in State, Figures Show." *Lexington Herald-Leader*. DIALOG File 721. July 24, 1991.

D47. Spevacek, Jennifer. "Toxic 500 Behind Three-Fourths of Pollution." *The Washington Times*. DIALOG File 717. August 11, 1989.

D48. "The TRI: Two Views." *Louisiana Industry Environmental Alert*. The Gale Group. DIALOG File 16. May, 1993.

D49. "Thinking Small About Pollution; A 1992 Overview of Waste Minimization." *Chemical Business*. The Gale Group. DIALOG File 16. June, 1992.

D50. "Three Suppliers Top Citizen Group's Polluter List." *Rubber and Plastics News*. The Gale Group. DIALOG File 16. July 22, 1991.

D51. "Top Ozone-Depleting Emissions, State by State." *USA Today*. DIALOG File 703. January 17, 1990.

D52. "TRI, TRI Again." *Louisiana Industry Environmental Advisor*. The Gale Group. DIALOG File 16. March, 1994.

D53. Walters, Rebecca. "Rural County Climbs to No. 4 on State's Toxic List." *Business First-Columbus*. The Gale Group. DIALOG File 148. April 13, 1992.

D54. Weir, David and Yamin, Priscilla. "Toxic Ten: America's Truant Corporations." *Mother Jones*. The Gale Group. DIALOG File 88. Jan. – Feb., 1993.

D55. Graham, Mary; *Regulation by Shaming*; The Atlantic Monthly, April 2000.

# APPENDIX E

# ANALYSIS OF ACCIDENT DATA

# APPENDIX E

# ANALYSIS OF ACCIDENT DATA

## RMP Data

Facilities subject to the RMP rule are required to submit, as part of their RMP, information on all serious accidents that occurred in the five years prior to the date of RMP submission. Because RMPs were submitted from March through the end of 1999, both 1994 and 1999 represent partial year data. The table below shows the industry sectors that reported accidents in their RMPs; these sectors are chemical manufacturers (NAICS 325); petroleum refineries (NAICS 32411); facilities that use ammonia for cold storage (food processors, food distributors, refrigerated warehouses, and food warehouses, NAICS 311, 4224, 49312, 49313); pulp and paper mills (NAICS 3221); chemical wholesalers (NAICS 42269); drinking water treatment plants (NAICS 22131); and wastewater treatment plants (NAICS 22132).

## Table E-1 – RMP Accidents Reported by Industry Sector

| Sector: | 1994 | 1995 | 1996 | 1997 | 1998 | 1999 | 5 Year TOTAL | Annual Average* |
|---|---|---|---|---|---|---|---|---|
| Chemical Manufacturers | 63 | 104 | 128 | 145 | 126 | 52 | 618 | 126 |
| Refineries | 17 | 31 | 50 | 34 | 38 | 23 | 193 | 39 |
| Cold Storage | 30 | 64 | 80 | 92 | 96 | 29 | 391 | 78 |
| Pulp and Paper | 6 | 28 | 24 | 20 | 22 | 7 | 107 | 21 |
| Chemical Wholesaler | 5 | 11 | 16 | 27 | 22 | 6 | 87 | 17 |
| Water Treatment | 11 | 24 | 18 | 20 | 29 | 14 | 116 | 23 |
| POTWs** | 12 | 19 | 22 | 24 | 24 | 9 | 110 | 22 |

\* 5 year Total divided by 5; assumes part year data for '94 and '99 are together equivalent to a full year.
\*\* Publicly Owned Treatment Works for waste-water treatment

## ERNS Data

The Emergency Response Notification System (ERNS) is a database used to store information on notifications of oil discharges and hazardous substances releases. The ERNS program is a cooperative data sharing effort among the Environmental Protection Agency, the Department of Transportation, and the National Response Center (NRC). ERNS provides the most comprehensive data compiled on notifications of oil discharges and hazardous substance releases in the United States. Since its inception in 1986, more than 275,000 release notifications have been entered into ERNS.

ERNS releases for four states — MA, CT, NJ, and VA — were reviewed. The states were selected because of their similarity in size and industrial sectors. Releases were divided into three categories:

- Fixed facility hazardous substance releases. Releases from unknown sources, those generated by private citizens, and releases of unknown materials were deleted. Releases of building products, such as asbestos and asphalt, were also eliminated.

- Transportation releases, including all releases classified as transportation even if the source or substance was unknown.

- All releases of oil and oil products. Non-petroleum oil releases were deleted (e.g., mineral oil); releases listed as unknown oil were included.

Unless they were classified as transportation releases, a number of releases reported to ERNS were eliminated and do not appear in any of the analyses. These releases were of unknown substances (either listed as unknown material or unknown chemical or by vague descriptions, such as "smoky debacle," "green glowing stick," "chemical odor," and "smells like cat urine") or of substances clearly not subject to reporting (household appliances, used cars, injured duck, and tires). Many releases appear in both transportation and oil releases (petroleum products may represent up to 75 percent of transportation releases).

Hazardous substance releases were further analyzed as follows:

- Releases of currently listed TRI chemicals from facilities that appear to be manufacturers. Where the type of business was unclear, but could be a manufacturer, it was included.

- Releases of currently listed hazardous substances (including those listed by category) where the reported quantity is greater than the current reportable quantity.

Many of the reported releases of TRI chemicals are not reportable under CERCLA; this is particularly true of the CFCs (e.g., freon) and CFC substitutes.

**Table E-2 – All Hazardous Substance Releases**

| Year: | CT | MA | VA | NJ |
|---|---|---|---|---|
| 90 | 61 | 94 | 104 | 202 |
| 91 | 70 | 106 | 116 | 175 |
| 92 | 109 | 145 | 118 | 124 |
| 93 | 124 | 150 | 134 | 80 |
| 94 | 113 | 128 | 127 | 102 |
| 95 | 48 | 73 | 83 | 75 |
| 96 | 52 | 55 | 76 | 79 |
| 97 | 25 | 55 | 49 | 59 |
| 98 | 25 | 58 | 73 | 49 |
| 99 | 21 | 44 | 85 | 49 |

**Table E-3 – Releases of Current TRI Chemicals from Manufacturers**

| Year: | CT | MA | VA | NJ |
|---|---|---|---|---|
| 90 | 38 | 45 | 36 | 151 |
| 91 | 43 | 43 | 46 | 82 |
| 92 | 57 | 39 | 42 | 61 |
| 93 | 56 | 51 | 49 | 44 |
| 94 | 63 | 42 | 55 | 44 |
| 95 | 26 | 32 | 24 | 28 |
| 96 | 13 | 21 | 33 | 30 |
| 97 | 12 | 15 | 22 | 32 |
| 98 | 9 | 16 | 24 | 22 |
| 99 | 11 | 17 | 29 | 28 |

**Table E-4 – Releases of Hazardous Substances Above the Current Reportable Quantity**

| Year: | MA | CT | VA | NJ |
|---|---|---|---|---|
| 90 | 23 | 29 | 58 | 100 |
| 91 | 30 | 42 | 66 | 53 |
| 92 | 34 | 56 | 32 | 53 |
| 93 | 27 | 29 | 63 | 25 |
| 94 | 23 | 30 | 41 | 33 |
| 95 | 21 | 21 | 35 | 21 |
| 96 | 13 | 14 | 26 | 22 |
| 97 | 13 | 7 | 17 | 27 |
| 98 | 5 | 10 | 29 | 20 |
| 99 | 7 | 6 | 41 | 13 |

**Table E-5 – Total Hazardous Substance Releases for Four States**

| Year: | All Hazardous Substance Releases | TRI Manufacturer Releases | Reportable Releases |
|---|---|---|---|
| 90 | 461 | 270 | 210 |
| 91 | 467 | 214 | 191 |
| 92 | 496 | 199 | 175 |
| 93 | 488 | 200 | 144 |
| 94 | 470 | 204 | 127 |
| 95 | 279 | 110 | 98 |
| 96 | 262 | 97 | 75 |
| 97 | 188 | 81 | 64 |
| 98 | 205 | 71 | 64 |
| 99 | 199 | 85 | 67 |
|  |  |  |  |
| 1999/peak | 40% | 31% | 32% |

## Table E-6 – Oil Releases

| Year: | MA | CT | VA | NJ |
|---|---|---|---|---|
| 90 | 612 | 113 | 535 | 503 |
| 91 | 700 | 128 | 510 | 518 |
| 92 | 701 | 151 | 806 | 483 |
| 93 | 616 | 184 | 822 | 479 |
| 94 | 669 | 281 | 876 | 645 |
| 95 | 364 | 194 | 743 | 629 |
| 96 | 539 | 359 | 717 | 659 |
| 97 | 447 | 246 | 527 | 418 |
| 98 | 473 | 271 | 794 | 415 |
| 99 | 368 | 260 | 754 | 442 |

## Table E-7 – Transportation Releases

| Year: | MA | CT | VA | NJ |
|---|---|---|---|---|
| 90 | 165 | 58 | 285 | 295 |
| 91 | 182 | 52 | 384 | 293 |
| 92 | 208 | 52 | 295 | 253 |
| 93 | 192 | 81 | 434 | 231 |
| 94 | 215 | 117 | 493 | 312 |
| 95 | 115 | 86 | 405 | 297 |
| 96 | 136 | 115 | 450 | 314 |
| 97 | 110 | 93 | 355 | 189 |
| 98 | 195 | 87 | 465 | 214 |
| 99 | 104 | 113 | 450 | 239 |

# APPENDIX F

## INSTITUTIONAL USES OF OCA DATA
## CREATING INCENTIVES FOR RISK REDUCTION

# APPENDIX F

# INSTITUTIONAL USES OF OCA DATA
# CREATING INCENTIVES FOR RISK REDUCTION

The power of public scrutiny can manifest itself in many ways. Research reveals that many types of institutions and individuals have interests that can be served by accessing Off-site Consequence Analysis (OCA) information, as noted in **Chapter 3**. This appendix examines segments of the public, their interests, their needs for access to OCA information, and their likely uses of it. The nature of the access desired is also discussed.

Segments of the public discussed include:

- • Residents and community, public interest, and environmental organizations
- • News media
- • Emergency planning and response organizations
- • Industry and trade associations
- • Those with a financial interest in the company
- • Local officials and major local financial stakeholders
- • Workers and labor unions
- • Universities and research organizations
- • Professional organizations

Please note the following. An interested party may act with risk reduction as an explicit goal, *or* it may act out of *other* interests (e.g., a newspaper seeking to improve its circulation) and still lead to risk reduction. Also, except where noted, the parties' access described here is as members of the public, rather than as covered persons or qualified researchers under the Chemical Safety Information, Site Security and Fuels Regulatory Relief Act (CSISSFRRA).

**Residents and Community, Public Interest, and Environmental Organizations**

> Residents and non-governmental organizations have a keen interest in access to OCA information. The greater the disclosure, the more likely it is that citizens will engage in dialogue with companies, governments, and their families to address chemical hazards. In addition, there is ample evidence that for information to be used, it must be easily accessed. Local residents and local groups are most interested in information about facilities nearby. More broadly-based organizations need the ability to compare the full range of facilities across the nation.

In 1992, a non-governmental organization in West Virginia's Kanawha Valley, along with the National Institute for Chemical Studies, petitioned local chemical facilities to disclose worst case accident scenarios. Two years later area chemical companies voluntarily did so at a public event held where the most residents would normally be found – in a shopping mall. Although there were concerns about panic or strong negative reaction in the community, these did not occur. Instead, the effort led to a number of positive outcomes.[1] Unfortunately, although the Kanawha Valley experience was positive, such voluntary disclosures are rare, and generally will not serve the needs of the vast majority of communities.

Wide disclosure of OCA information would inform citizens and citizens' groups, which generally do not have access as covered persons under the statute. If the information raised their concerns, it would stimulate activity. For example, they would more likely register concerns with company management and request risk-reducing steps be taken. Likewise, they would more likely contact local and state officials to register concerns and request that the officials take steps – from requesting voluntary action of a facility to changing its operating permit. On an individual and family basis, informed residents can take self-protective actions, such as learning about emergency response and sheltering-in-place, participating in emergency drills, or moving residences or children's daycare and school locations. Note that many of these actions can reduce the risk posed by a chemical release, *no matterwhether accidental or not.*

In addition, concerned citizens may also seek information about facilities in other counties or states to compare their situation with that of friends, relatives, or others. Real estate guides now recommend that families find out environmental information about particular neighborhoods on the Internet before they relocate.[2] Ideally, citizens and organizers could examine the OCA information (both worst-case and alternative release scenarios), along with the complete RMPs, for other facilities and learn about their potential impacts, mitigation measures, and prevention programs. The availability of this information would likely stimulate dialogue with local facilities about differences in programs from one facility to another. Further, citizens may want to learn about emergency response measures and capabilities at other communities associated with alternative release scenarios to push for improvements in their own communities.

Ease of access has been shown to be vital for citizens and non-governmental organizations. Environmental data sources that are easy to use see heavy usage, while those that are harder to access do not. From July 1999 through January 2000, EPA's RMP*Info on the Internet, which includes the RMP Executive Summaries, has logged over 155,000 page views of RMPs. Note that RMP*Info does not currently provide OCA information; with OCA information the usage would likely be higher. The "Scorecard" website from Environmental Defense (formerly Environmental Defense Fund) likewise has found a high level of use. Scorecard provides data on releases of Toxics Release Inventory (TRI) and other pollutants and data on ambient air quality, interpreted for health risks and mapped for geographic sorting and display. Scorecard has hosted over two million visitors since its launch in April 1998, and serves about 600,000 page views a month. The website also offers the ability for visitors to send a fax directly to a company; about 5,000 faxes have been sent to about 3,000 distinct companies.[3] TRI data, easy to obtain on the Internet, have been used extensively by public interest and environmental groups, with the three most frequently reported uses being directly pressuring facilities for change, educating citizens, and lobbying for policy changes.[4]

In contrast to the high usage of Internet websites, the public rarely accesses risk-related data available only at state and local agencies that are not on the Internet. Information on hazardous chemicals present at facilities and their quantities is provided to state and local agencies under the Emergency Planning and Community Right-to-Know Act sections 311 and 312. In one study, 41% of local emergency planning commissions (LEPCs) had no public inquiries during the year of the study, and only 25% received over six inquiries.[5]

This low level of information transfer applies also to risk information in states where facilities file reports, but where the information is not made available to the public except on request. For example, New Jersey Bureau of Chemical Release Information and Prevention has gathered information similar to RMPs for a number of years and makes paper copies available to those who travel to its office or ask for them by mail; however, it has received fewer than 10 requests.[6] Delaware has received similarly few requests.[7] Nevada places its reports in public libraries, but its experience indicates that librarians have to hunt for the files when updates are needed. Few people have requested information from the state unless there has been an accident.[8] Access is low, although these three states have had accident prevention program rules in place since the late 1980s or early 1990s.

Why the disparity described above? The difference in usage is explained in part by what researchers call the collective action problem. This problem occurs when a person chooses not to take an action that would have net benefits to society because the costs to the individual of taking the action outweigh the benefits to that person, or because he or she expects someone else to take the action. Public information mitigates this problem by lowering these costs.[9] Interest may be substantial, but if access requires traveling across the county, requesting a specific document by mail, or calling during the work day, many citizens will be unable or unwilling to gain access. Making information publicly available helps overcome this problem.

Community and environmental organizations have expressed interest in comparing facilities within a sector, within a region, from state to state, or within a company. For example, in one case a refinery was up for sale and an environmental group sought to learn about the environmental performance of companies that were potential purchasers. Non-governmental organizations want to use data from OCAs to evaluate a facility's or company's hazard – in both absolute terms and also, with national access to relevant data, in relative terms.[10]

The issue of public disclosure of hazard information was raised recently in Sacramento, California. There, after the FBI arrested two individuals in an alleged plot to attempt to blow up a large propane storage facility, the press reported on area residents' concerns. According to one of multiple reports, "Subdivision residents plan to meet to discuss getting the propane tanks moved or security increased, said Karen Banda, who bought a house there with her husband two years ago. They asked the developers about the plant at that time, she said. 'If I remember correctly, what they said was there was no concern,' Ms. Banda said. 'They should not have been allowed to build [housing] this close.'"[11] Meanwhile, the facility's own RMP describes a far less damaging worst-case scenario than reported in some accounts. The company's OCA yields a worst-case distance to endpoint of 0.50 miles, based on one pound per square inch overpressurization.[12] Not only does this case show that OCA data are not necessary for would-be criminals to target facilities with toxic or flammable chemicals (the facility made OCA data available only later), but it reveals the value of public hazard information. Had this facility been part of the RMP program (as it is now), and had the OCA information been easily accessible, then residents – and potential residents – could have known what hazards existed in the area and acted based on their level of concern.

Hazard information can raise concerns or it can lower them. Researchers at the Rutgers University found that comprehensive outreach materials on accidental release scenarios not only educates the public about risks and proper accident response, but it may actually serve to increase trust and decrease worry.[13] It is true that some facilities have voluntarily shared the OCA portions of their RMPs and will continue to do so, and that OCA information does not address the likelihood of a release. However, if members of the public could add widely available OCA information to their understanding of accident likelihood (informed in part by RMP accident histories), then the level of their concern likely would be in line with the apparent risks.

Equity is also a relevant consideration. A drawback of restricting access is that it will likely have a disproportionate impact on poor and minority communities. RMP facilities may be more often sited in poor, minority, or immigrant communities. If information is difficult to acquire and interpret, then those with fewer advantages in income, language, or education are less likely to be able to understand the hazards they face than those with more advantages. Traveling to government offices, using raw data to construct OCAs, or hiring consultants to ferret out and analyze highly technical information are all substantial hurdles. OCA information, since it is interpreted and provides relevant information such as the distance that harm could potentially extend, acts to overcome these hurdles. In addition, news media, citizen groups, and facilities can bring OCA information to the people, counteracting to some degree the disparities in information handling capacity. Granted, relying solely on the Internet for disclosure would also have disproportionate impacts, at least in the short run. However, while a significant portion of the public, including many in disadvantaged groups, do not have access to the Internet at home, many gain access at work or at school, and the portion of the public with ready access is growing rapidly.[14] Other means, such as mail-out hard copies of OCA information, could be used to address information needs for the time being.

What is the nature of access that would be used by citizens and citizens' groups for risk reduction? Certainly, the public located near facilities handling hazardous substances will be most interested in the OCA information and other RMP data for those local facilities. The statute requires that the regulation provide for public access to a "limited number" of paper copies. However, some localities have a large number of RMP facilities, and relying on paper copies and read-only access means less use of OCA information.

Nature of Access That Would Be Useful

For facilities in the local area, all OCA data elements, as well as the rest of an RMP, would be valuable in better identifying and understanding local hazards. In addition, local residents and local organizations desire information from OCAs for facilities with similar processes or chemicals throughout the country, in order to allow comparisons. If someone were to process OCA data into aggregations that happened to correspond to the needs of those living near a facility and provide it to them, that facility could be compared with high, low, and average values for a particular grouping, without facility identification. However, the lack of identification information would prevent dialogue with facilities. Full OCA information (all OCA data elements in a computerized database) would most efficiently serve the need for drawing comparisons. Finally, national and state environmental and public interest organizations would be likely to use all elements of OCA information for all facilities, for the same reason of allowing comparisons and contacting facilities and companies of concern.

**News Media**

> News media organizations, proven to be a large factor in the Toxics Release Inventory reductions, have shown their interest in reporting on chemical hazards. Were it more publicly available, the media would likely use OCA information to gain easier access and obtain accurate, consistent information set in context, on facilities both nearby and across the country. By reporting on OCA information, the media will stimulate risk reduction by focusing the attention of actors (communities, industry, and government) on the potential off-site consequences of chemical releases and ways to prevent and mitigate them.

Newspapers and other media have a strong interest and a long tradition in informing the public about risks that they face, and what can be done about the risks. Importantly, coverage of chemical hazards can raise awareness and motivation among many of the other parties discussed in this chapter. This coverage stimulates action by stakeholders to engage in dialogue and reduce risks.

Ample evidence supports the news media's interest and power regarding issues of chemical release risk. Several papers have carried articles highlighting the RMPs (without OCA information) and hazards of local facilities. One example of press attention is a recent series by the *Washington Post*. Following an article describing worst case scenarios of several local facilities, the paper focused on concerns about the handling of chlorine at the Blue Plains sewage treatment plant in Washington, DC. The plant had voluntarily put detailed OCA-related information in its RMP Executive Summary, which is available widely. The paper described scenario details (including a map showing the area potentially at risk), the health effects of the chlorine and sulfur dioxide used at the plant, and apparent problems with plant safeguards. There were subsequent articles over the next several days, and city officials, including the head of local water authority and the mayor, took actions to improve safety monitoring, replace chlorine one and a half years earlier than previously planned, and beef up security. (Comparisons to OCAs of similar facilities outside the region were not published.) [15]

The media's effect in the Risk Management Program would likely resemble that in the TRI program. There, a majority of respondents from citizen groups, state TRI agencies, and industry agreed that media coverage of toxics issues has increased since the implementation of TRI.[16] Newspapers have run articles listing specific companies or facilities as the top polluters in a town, a State, or the country. Not only has coverage increased, but negative coverage has corresponded with dramatic reductions in TRI releases; **Appendix D** provides a study of this effect.

Newspapers and other media organizations have limited time and resources to research stories; if easily accessible, OCA data likely would be used and chemical hazards reported more completely. If, on the other hand, media organizations must obtain raw data from multiple sources and then compile and analyze the resulting information about OCAs, the media would likely report less. In addition, OCA data, as interpreted data that is already calculated and put into a relevant context, would tend to be the most newsworthy part of RMPs. If it were available, OCA data would likely attract the media's and the public's attention, first to the OCA data, and then to other RMP elements.

News organizations have shown that they are willing and able to report on chemical release risks, and this reporting clearly brings about risk reduction over time. However, while the media are to some extent able to pursue their interests in hazard information locally, were searchable OCA data available on a national basis, the media could more easily put hazards into understandable terms for the public and draw comparisons among localities and facilities.

<u>Nature of Access That Would Be Useful</u>

In general, the news media are interested in accessing all OCA information on a national basis. Certainly, local news organizations are most interested in information about local facilities. However, access to all data elements and the ability to compare facilities with one another would greatly enhance the ability of national and local media to put chemical hazards into context for the public.

**Emergency Planning and Response Organizations**

While local and state organizations responsible for emergency planning or response are and will be entitled to OCA information, greater public disclosure carries advantages for them and for risk reduction. These agencies often lack resources to fulfill their planning tasks and often are concerned about penalties for improper dissemination of OCA information. The greater the public accessibility to OCA data, the less the public must rely on these agencies for its hazard information. Moreover, public access may ease the planning and response burden for these entities because access will promote better reporting by companies and more participation in planning and response by all stakeholders.

OCA information and RMPs in general detail potential harm to the community (and to responders). Planners and responders have responsibility to mitigate and reduce chemical release risks to the community and to responders.

Local government officials will have access to OCA information as "covered persons" for official use under CSISSFRRA. However, local organizations will often need some way to gauge the level of risk and to evaluate the nature of their emergency preparedness and response. What better way to gauge this effort than to use OCA information (worst-case scenario to some extent, but alternative release scenario even more so) to identify locations that have similar problems in order to learn about improvements in risk management or emergency preparedness and response? Without broad OCA information availability, there will be no practical way to learn about the passive and active mitigation measures used with the OCA assessments, and seeking out relevant communities and facilities will be hit or miss.

Although state and local agencies are allowed access to OCA information for official use as "covered persons" under the statute, there are two factors that restrict the further sharing of information through these agencies. The first factor is that of resources. No stream of federal funding is provided for LEPCs or SERCs, and they are often quite short of resources. In the Mountain West, an estimated 75% of LEPCs have resource issues.[17]

Due partly to the resources problem, emergency planning and response agencies can benefit, and information transfer can be enhanced, if information is available from a variety of sources aside from themselves. For one, to the extent that hazard data could be obtained and transferred without requiring local or state resources, the existing strain on resources is not increased. In addition, broad dissemination allows public comparisons in a public forum, versus dialogue only between facilities and covered persons. Broader dissemination would tend to bring in outside viewpoints and issues and reduce the reliance on what are in fact often volunteer organizations.

Regarding LEPCs' resources, some of the reasons public disclosure is beneficial have to do with LEPCs' strengths and weaknesses. In general, LEPCs have not made a concerted effort to bring hazardous materials issues to public attention, focusing instead on technical aspects.

Furthermore, given the constraints under which LEPCs operate, it is unrealistic to expect LEPCs to attempt to foster public debate of environmental issues or to focus on hazard reduction rather than emergency response.[18] A Government Accounting Office report noted that EPCRA section 312 chemical inventory data, which are provided to SERCs and LEPCs, are usually not even computerized, and that use of the data by the broad public has been limited.[19] Due to the resource constraints, having information available from multiple reliable sources would reduce the reliance on LEPC officials (and other covered persons), who otherwise would be responsible for responding to more requests and questions with their limited resources. This, in turn, would lead to both a reduced strain on agencies and greater public access to data of interest.

The second factor restricting information flow through state and local covered persons is liability. CSISSFRRA includes a prohibition on unauthorized disclosure of OCA information by Covered Persons. This prohibition, and the potential criminal penalties and fines of up to $1 million for violations judged to be willful, pose large disincentives for local and state officials to pass on information involving OCAs. Many officials perceive that they are liable, and do not want to be the arbiter of who gets access to what information. The statute continues to have a chilling effect on officials obtaining and distributing information related to release risks.[20]

Government officials would likely make beneficial use of publicly available OCA information, as they have other widely available information. For example, environmental agency heads have publicly called upon the firms with the highest TRI emissions to voluntarily reduce their releases. In addition, for the state agencies that run TRI programs, the three most frequently reported uses of TRI data are comparing the data to permits, source reduction efforts, and comparing emissions patterns at similar facilities.[21] Note that comprehensive comparisons are possible only with access to nationwide data.

Responders and planners benefit indirectly from public access to the extent access promotes compliance by industry and public involvement in planning. The additional scrutiny of RMP compliance resulting from public access will lead to more complete compliance and higher quality RMPs. This will, in turn, allow planners to use on these inputs and simplify planning. Greater public involvement in planning will give planners more leverage in dealing with risks in the community.

The drawbacks of restricting OCA information access and distribution include the following. Local planners, fire departments, and others currently do not want to obtain OCA information, in order to avoid liability if the same data were somehow released while in their possession. Some local planners have indicated that because of the severe ($1 million) potential penalty, they would rather not take possession of the data, regardless of whether they are entitled or have access.[22] The perceived negative impacts have generated a chilling effect on the desirability and use of OCA information and even other information associated with it.[23]

<u>Nature of Access That Would Be Useful</u>

The types of access needed by emergency planning and response organizations are similar to those that would be useful for citizens and citizens' groups. In general, for facilities within an agency's area of responsibility, nothing less than all OCA information makes sense, although alternative release scenario information is perhaps most useful to planners. This means that local agencies would likely make use of local information and state agencies would use statewide information. However, for facilities with similar processes, both local and state agencies can benefit from disclosure of OCA information from such facilities nationwide.

## Industry and Trade Associations

> Companies with facilities subject to RMP requirements, which have the ability to directly reduce the risk of chemical releases, have strong incentives to use OCA information, including that from other companies, to efficiently reduce and manage their risk. Producing high quality OCAs and accessing OCA data from other companies would be most effectively achieved with a publicly available database.

This subsection describes the major incentives for risk reduction related to industries with chemical release risks and their trade associations stemming from increased accessibility to OCA information in a readily accessible form. In addition, this subsection discusses potential consequences of restrictions of access to OCA information on industry's ability to prevent catastrophic accidents.

The major categories of benefits or incentives for industry and trade associations associated with the widespread dissemination of OCA information are:

- General Duty
- Potential for Greater Risk Reduction
- The Need for Continuous Improvement, and Demonstrating a Capability to Manage Risk

Each of these categories are briefly described below.

### General Duty

One of the most important prevention-related sections of the CAA is section 112(r)(1), the General Duty Clause. This section provides that industry has a general duty – an obligation to understand the hazards of their operations, design and operate a safe plant, take the necessary steps to prevent accidents, and act to mitigate the consequences of those releases that do occur. Consequently, satisfaction of this general duty means that a facility owner or operator must be fully aware of all of the hazards present at its facility. In addition, the owner or operator must be knowledgeable about, and make use of, hazard information, prevention practices, emergency response practices, and all relevant industry codes and standards and regulations that apply to the industry. Full and complete RMPs and OCA information are essential tools that must be used to the extent possible by industry as required by the general duty, and broad dissemination makes using these tools more practical.

Potential for Greater Risk Reduction

Two key aspects are discussed: OCA information's critical role in process hazards analysis; and how OCA information leads to risk management action.

At the heart of chemical accident prevention and process safety management is a formal process hazards analysis, or PHA. A PHA involves the regular evaluation and identification of hazards, assessment of risk and selection of risk control alternatives throughout the operating lifetime of a facility. The majority of relevant information needed for a PHA is based on the processes used at the facility. However, OCA information along with complete RMP details assembled in one location for immediate use by hazard evaluation analysts across a number of different industries and processes would broaden the process safety knowledge applied to PHAs. A greater understanding of the wide variety of hazards, accident impacts, mitigation measures and prevention programs serves to reduce the potential for catastrophic chemical accidents in specific processes. The importance of various elements in a chemical accident prevention program is often driven by the potential consequences of process upsets or failures. These consequences are mirrored in the OCA information, especially in the alternative release scenario. Consequently, suppressing OCA information from widespread publication would tend to diminish the overall value and context of the RMP prevention information available to all industry and increase the chance that information useful for a PHA was unavailable.

A thorough PHA along with hazard and consequence assessments is designed to reveal vulnerabilities in a system that could lead to disaster. Once these vulnerabilities are revealed, changes can be made to the chemicals, process technology or safeguards to address the vulnerability. This leads to inherently safer processes.

The Risk Management Program and Risk Management Plan regulatory requirements are flexible regarding facilities' equipment and operations. Although the rules provide a specific framework, companies are obligated to assess for themselves the chemical and process hazards and off-site consequences present at the site and to develop integrated accident prevention measures and emergency response plans tailored specifically to the needs of the facility. The facility bears the responsibility for evaluation, implementation, and documentation in the risk management plan of the measures that will be used to protect the public and environment. With this responsibility comes the burden of obtaining as much information as possible so that these tasks are carried out as effectively and as accurately as possible. Widespread publication of the plan *for all to see* drives a facility to make efforts that are more effective than under traditional command and control regulatory approaches dictated by government. Only with publication of complete RMPs along with OCA information, ideally in computerized form, can full use of RMP information be made. Disclosure short of this would lead to reduced levels of effort.

In order to gauge the degree of consequences associated with their worst case scenarios relative to their peers, companies must be able to compare their results with those of other facilities. This could be done to some extent by using databases without identification information or summaries showing data value ranges and averages (assuming someone would compile such summaries). However, only with full OCA information in database form can a company not only compare its performance with that of others, but use the identifying information to contact its peers. Thus, publication of OCA information is shown to be important for hazards analysis.

The second important aspect of potential risk reduction is that OCA information leads to risk management action.

EPA expects industry to react to public disclosure of OCA information in ways similar to its reaction to publication of the Toxics Release Inventory. Industry reacted in very significant ways to TRI. As related in **Chapter 3**, TRI emissions have decreased by 43 percent since 1988, due to multiple factors. Case studies of eight companies and their response to the Emergency Planning and Community Right-to-Know Act found that EPCRA "reporting and emergency planning requirements have advanced the internal level of company awareness of their chemical-related activities and releases, leading to the identification of areas for improvement and subsequent action...[EPCRA] has resulted in greater transparency of company activities and [this] has led to corporate action to reduce chemical risks."[24] The two most frequently reported uses of TRI data by industry are source reduction efforts and educating citizens.[25] In one well-known example, at the time of distribution of the initial TRI results, the Chief Executive of Monsanto committed the company to a 90% reduction in TRI releases.

The evidence goes on. In one random survey of industry representatives, most respondents reported that they were much more concerned with toxic releases since the passing of EPCRA. The survey found a correlation between industries that said they had publicized environmental problems or visible pollution and those industries that increased risk communication with the public.[26]

Companies have increasingly joined voluntary programs intended to bring about change without regulatory mandates but with public information. One example is EPA's 33/50 Program. Launched in 1991, the program's goals included large reductions in releases of 17 TRI chemicals from all reporting facilities. Measured relative to a 1988 baseline, the goals were for 33% reduction by 1992 and 50% reduction by 1995. Although the exact extent to which the 33/50 Program caused changes in behavior is uncertain, 13,000 companies did participate and the reduction goals were met ahead of schedule.[27]

Although there are differences in TRI and OCA information, EPA believes that the disclosure of OCA information would lead to many risk management actions and a downward trend in accidents and consequences similar to that of TRI releases.

The Need for Continuous Improvement, and Demonstrating a Capability to Manage Risk

A principle associated with good process safety management is the enhancement of process safety knowledge and continuous improvement.[28] Enhancement of process safety knowledge involves internal and external research to "benefit from the latest advances in process safety technology, and keep abreast of technological advances ..." (Ibid).

Even DuPont, long recognized as a world leader in safety, indicated that it has learned from dialogue with citizens. "When a community resident and Citizen's Advisory Panel member toured a DuPont facility after a recent release, [the member] suggested adding an ammonia monitor at the top of a storage tank. The suggestion was judged to be a good one and DuPont has since installed the device on the top of the tank."[29] Clearly this example demonstrates the value of local dialogue. However, the RMP and its related OCA information, data, accident histories, and prevention measures provide a basis for this kind of dialogue nationwide.

Under Responsible Care®, member chemical companies are to: "continuously improve health, safety, and environmental performance; listen to and respond to public concerns; and assist each other to achieve optimum performance."[30] (Responsible Care® is a chemical industry initiative that is built around a set of six Codes of Management Practices. These codes include Community Awareness and Emergency Response and Process Safety. Other elements of Responsible Care®, which companies may choose to join, include a self-evaluation process that determines how well companies are applying the Codes, mutual assistance, and performance improvement measurements. See www.cmahq.com and www.socma.com/respcare.) Indeed, changes at some facilities have been stimulated by a sharing of OCA information. In addition, complete RMPs containing OCA information can be useful tools to identify facilities that are safety leaders. Furthermore, small businesses with little technical resources and companies evaluating opportunities for new business development can learn from current industry practices to better control their future risks. Finally, chemical industry companies that are under Responsible Care® "encourage and help other chemical manufacturers ... improve their own performance through the Mutual Assistance Network, which provides direct interaction between companies and the Partnership Program, which allows non-members to participate in Responsible Care."[31]

Recently the chemical industry has worked to improve its public image. Part of this openness is driven by Responsible Care® as noted above. In addition, companies have found that dialogue is good for business. Certainly, making OCA information readily available for all to use has image and marketing advantages for many businesses, and can be a good starting point for dialogue with stakeholders. Many companies value their image as a responsible neighbor, and realize that negative public relations can hamper efforts to expand operations or renew permits.

Note that Responsible Care® is specific to the chemical manufacturing industry. Although Responsible Care® has been criticized as lacking in measurable goals, timelier, accountability and credibility[32], the many sectors that lack such an initiative may need even more help in addressing risks. Access to complete RMPs (containing OCA information) would broaden the network of assistance to the thousands of facilities that may not manufacture chemicals, but do handle them.

Over time, companies would likely reduce the risk of releases by taking action due to concerns brought to light by OCA information. These actions would reduce release hazards, the likelihood of a release, or both. Note that many risk-reducing actions, particularly those that reduce hazards, would reduce the risk of *accidental* releases *and* any risk that might be posed by *non-accidental* releases. Examples of these actions include material substitution, reduction of chemical inventories, passive mitigation measures, and emergency planning.

Some drawbacks to restricting OCA data access have been mentioned earlier in this subsection. In addition, restrictions may allow important data to be withheld even from facilities' neighbors, limiting the capability of local citizens to participate in risk management and emergency response planning. Those interested in OCA information would likely find a "work-around," and use other data and readily available tools (such as EPA's tool, RMP*Comp) to calculate and publish their own versions of worst-case and alternative case scenarios instead of the facilities' own. Such results are likely to not reflect conditions at the facility (e.g., process quantity, passive mitigation, specific weather conditions), leading to misleading results and driving attention to hazards that actually may be less significant and away from greater hazards.

<u>Nature of Access That Would Be Useful</u>

Companies and trade associations likely would make good use of several types of access to OCA data, as follows. Individual companies could benefit from information on similar processes from across the nation. Trade associations could make beneficial use of at least the same.

## Those with a Financial Interest in the Company

> Those with a direct stake in a company's success at preventing releases have an inherent interest in protecting that stake. Thus, to the extent OCA information is easily available and covers similar facilities, insurers and others can use this information to properly value their exposure and to pressure a company to improve where needed.

Disclosure of OCA information can stimulate concerns about potential economic impacts on those with a direct stake in a company with facilities handling hazardous chemicals. These parties include insurers, investors, lenders, and those with other business arrangements with a company. These interests are in addition to those of stakeholders in the *community*, described below.

Insurers are interested in the ways companies manage risk. In general, they factor indications of risk present into underwriting decisions and pricing. Regarding OCA information and RMPs, conversations with insurance and loss protection corporation representatives reveal that these companies would make significant use of this information. As with accident histories in RMPs, the details found in OCA information provide insurance and loss control companies with greater understanding of the kinds of scenarios and potential losses that could occur, validating the degree of loss control needed and validating expected insurance needs. Further, it would enable the loss control industry to address customers' needs for assistance.[33] Regarding process industry and process safety, "what industry needs is to continue the development of performance-based standards and regulations that follow the systemic approach and employ life-cycle models that include the process itself, all safety/control equipment, and people (operators and community). The approach must rely on risk metrics to support prudent business decisions ...", according to one leader in the insurance industry.[34] Disclosure of OCA information disclosure would support for such community considerations and standardized risk metrics.

Another interested sector is comprised of investors and investment advisors and researchers. These parties may disfavor purchase or ownership of company shares, and thereby cause reduction in market value of shares. Interest is keenest in the field of "socially responsible investing" (SRI). One of every eight dollars under investment management is now invested under the rapidly growing umbrella of SRI, where an asset is evaluated along both traditional financial measure and other measures.[35] Those involved in SRI are very interested in environmental data, and in comparing companies nationally.[36] The level of interest among non-SRI investors and portfolio managers lags behind.

Both mainstream and SRI fund managers often obtain data from research services. These services tend to use data available from national databases, such as the Securities and Exchange Commission's Electronic Data Gathering, Analysis, and Retrieval system (EDGAR) system. They generally do not use data at the state and lower levels, due to the lack of consistency from one data source to another, and due to the workload involved. Investment companies and research services generally support enhanced access to information related to risk, but often only at the federal level.[37]

Lenders comprise another interested group; they may be concerned that a potential economic loss from an accident threatens a company's ability to repay loans, and also threatens the value of assets (e.g., property) pledged as security for such loans. Other interested parties are firms seeking to acquire a particular company, merge with it, engage in a joint venture with it, or choose it as a primary supplier. They may be dissuaded from entering into such business arrangements. This could be due to a perception that the company could not perform due to an accident or its effects, or due to concern about incurring a share of accident liabilities under "joint and several liability" or other liability-sharing doctrines.

Some organizations with a financial stake in the company perhaps would apply for access to OCA information as qualified researchers. However, this would entail delay and restrictions, and thus will be used less frequently than if OCA information were publicly available.

It is clear that the parties described will have a stake not only in a company, but in its hazard information. In general these parties are likely to seek OCA information from companies with similar processes on a national basis.

## Local Officials and Major Local Financial Stakeholders

> Those with a stake in the community have an interest in the prevention of accidents, and would likely use OCA information to learn about the hazards posed by local facilities, and how those compare with those in other areas.

Local leaders who are not covered persons under the statute can benefit from public disclosure of OCA information. Such leaders include government officials not in emergency planning and response agencies (such as mayors) and persons with major financial stakes in the community (such as landowners, real estate developers, and members of the chamber of commerce). These interests are in addition to those of direct financial stakeholders in a particular *company*, described above.

Disclosure can reveal risks to community safety, property, infrastructure, and natural resources, and thereby prompt local leaders to use their powers and influence to prevent harms, including the political and economic repercussions of an accident. In addition to concern over risks stemming from an accident, there may be also be concern that information revealing large and persistent hazards will itself cause economic loss (such as driving down property values). Those affected by this would tend to take action to protect their interests.

The lack of knowledge about neighborhood hazards, as in the Sacramento case raised above, applies not only to residents, but also to other public officials. For example, a school board in Georgia hired a consultant to study the best locations for a new school building. Available sites were near industry. OCA data were requested from industry, but withheld. Consequently, data were derived from other sources and used in modeling; however a lack of complete information caused the results to be misleading, generating significant land use planning difficulties. Had the information been readily accessible, the costs and hassles that were caused could have been avoided.[38] As a practical matter, in any number of cases local entities will be unable to obtain OCA data.

These local government and business leaders would likely make use of access to all local OCA information. They may want to use data from outside their area to make comparisons, but such data would be of less interest.

## Workers and Labor Unions

Those that work at facilities have a keen interest in knowing what hazards surround them; OCA information would likely be used to raise awareness and reduce off-site and off-site risks.

Although each company prepares its own OCA data, in general the data will not be shared with all employees and contractors. Sharing of info across all industry would allow workers and their unions to be more knowledgeable about hazards, safety controls, risk management, etc. Unions have worked successfully to improve facilities' safety and environmental performance, as noted in **Appendix C**. In addition, if a company were to provide incorrect information in an OCA and it were made publicly available, those knowledgeable about the facility could see and point out errors so that they could be corrected. In multiple cases, EPA has had to work with companies to correct obvious errors in RMP submissions, but cannot identify less obvious errors.

Since *workers* tend to live near facilities, they also would share the same concerns and want the same data access as other local residents, noted above. *Unions* will tend to share the same interests as research organizations, below.

**Universities and Research Organizations**

Universities and research organizations, by virtue of their educational and research missions, have a natural interest in information that can lead to better understanding the risks associated with chemical releases. The RMP and OCA information are avenues to generate research around the world.

Multiple disciplines, including chemical engineering and risk communication, could be advanced by easily available OCA information. Some institutions are particularly interested. For example, the main mission of the Mary Kay O'Connor Process Safety Center, part of Texas A&M University, is to improve safety in the chemical process industry.

Those in research organizations could certainly apply to be qualified researchers under CSISSFRRA. However, certain aspects of this law will affect the ability of researchers to access OCA information and publish findings based on it. First, the statute requires a system for access; even the most efficient system will likely entail some delay. Second, the system "shall not allow the researcher to disseminate, or make available on the Internet, the [OCA] information, or any portion of the [OCA] information." (CSISSFRRA.) Although OCA information is narrowly defined, this restriction can hamper the ability of a researcher to publish or support findings. The likely effect of these factors will be less research into accident prevention and chemical security. Thus, more immediate access and fewer restrictions on usage relative to the statutory minimums are of interest to researchers, whether they are with a university, an industry trade association, or a union.

Universities and research organizations would likely make use of all OCA information data elements across the country.

## Professional Organizations

> A number of professional organizations would likely use OCA information to promote process safety.

Professional organizations have an interest in promoting process safety within their professions. One example of an interested professional organization is the Center for Chemical Process safety (CCPS), founded in 1985 following several chemical incidents. It is a division of the American Institute of Chemical Engineers. CCPS commits itself to developing engineering and management practices to prevent or mitigate the consequences of catastrophic events involving the release of chemicals and hydrocarbons that could harm employees, neighbors and the environment. Some areas of interest to CCPS sponsors include hazard and risk analysis, engineering design, operations and maintenance, information dissemination and process safety management.[39]

The CCPS has established a program to foster education about hazards analysis, process safety, and emergency response at the undergraduate chemical engineering level. The Safety in Chemical Engineering effort utilizes industry information in undergraduate chemical engineering curricula so that graduating chemical engineers are better prepared to work safely in the chemical processing industries. Course work includes problem sets and study of the concepts in hazard analysis, consequence modeling, hazard evaluation, risk management, and process safety management. This program could benefit from OCA information in conjunction with RMPs as part of academic training and research.

In addition, the CCPS has produced numerous guidelines for process safety which rely on industry information; future guidance could rely heavily on the hazard assessment and process safety data contained in RMPs and OCA information. Further, CCPS conducts research into technical problems and issues associated with hazards analysis and process safety.

Another example of interested professional organizations is the American Chemical Society. It seeks, in part, to promote the public's understanding of chemistry and the chemical sciences, and to foster communication and understanding among its members, the chemical industry, the government and the community in order to enhance the quality of scientific research, support economic progress, and insure public health & safety.[40]

The access needs for these and other professional organizations would match those of universities and research organizations. That is, they would be interested in the most data elements possible, and in the maximum geographic reach.

## Summary of Natures of Useful Access and Likely Impact on Risk Reduction

To characterize in even a very general way the different natures of access desired, one must consider multiple variables. One variable is the *geographic extent* of information accessed, which could range from the county to the country. Another variable describes the *facilities* of interest. Interest in OCA information could range from all facilities to only one company, sector, or process. The last variable covered here represents the *data elements* of interest. For example, one party might make use of access to all data elements of an OCA, while another is content without exact location or facility identification. For each segment of the public described in the preceding section, Table 1 very briefly and generally notes the nature of interests in access to OCA information.

### Table F-1 – Summary of the Natures of Interests in OCA Information

| Type of organization/person | Geographic Extent of Interest | Facilities of Interest | Data Elements of Interest |
|---|---|---|---|
| Residents and *Local* Community, Public Interest, and Environmental Organizations | Local | All in local area | All (Individuals less interested in raw data) |
|  | National | Those with similar chemicals or processes | All (Individuals less interested in raw data) |
| *National and State* Community, Public Interest, and Environmental Organizations | National | All | All |
| News Media | National | All | All |
| Emergency Planning and Response Organizations: | | | |
| •• LEPCs and Fire Departments | Local | All | All |
|  | National | Those with similar chemicals or processes | All |
| •• SERCs | State | All in state | All |
|  | National | Those with similar chemicals or processes | All |
| Industry and Trade Associations: | | | |
| •• Individual Companies | National | Those with similar chemicals or processes | All |
| •• Trade Groups | National | All | All |

| Type of organization/person | Geographic Extent of Interest | Facilities of Interest | Data Elements of Interest |
|---|---|---|---|
| Those with Financial Stake in Company | National | All | All |
| Local Officials and Major Local Financial Stakeholders | Primarily local | All in local area | Distance to endpoint and receptors |
| Workers and Labor Unions | Local | Facility where working | All (Individuals less interested in raw data) |
|  | National | Those with similar chemicals or processes | All (Individuals less interested in raw data) |
| Universities and Research Organizations | National | All | All |
| Professional Organizations | National | All | All (Especially mitigation used) |

Which segments of the public would bring about the most risk reduction, given disclosure? It is impossible to make predict for any particular case, but EPA believes that certain segments will have a greater overall impact than others. The segments that would have the highest impact on risk reduction are:

- •• Residents and community, public interest, and environmental organizations
- •• News media
- •• Emergency planning and response organizations
- •• Industry

Even though each company has access to its own OCA information now, industry and trade associations would also have a significant impact if there were widespread disclosure. The remaining segments of the public will likely have impacts, but they may not be as great as those just mentioned. However, in any particular situation *any* type of person or organization could act to significantly reduce risk.

# APPENDIX F REFERENCES

F1. *NICS News.* Vol. 8. (Fall 1999): p. 6.

F2. Harney, Kenneth R. "Web Sites Offer Helpful Information to Buyers and Owners." Washington Post. (December 4, 1999): p. G12.

F3. EPA CEPPO; Correspondence with Dr. William S. Pease, Director, Internet Projects, Environmental Defense Fund. Nov. 22, 1999.

F4. Lynn, F. and Kartez, J. "Environmental Democracy in Action: The Toxics Release Inventory," Environmental Management, Vol. 18, No. 4, p. 511-521.

F5. Adams, William C., Stephen D. Burns, Philip G. Handwerk. *Nationwide LEPC Survey: Summary Report*. The George Washington University (Oct. 1994).

F6. Baldini, Reggie. New Jersey Bureau of Chemical Release Information and Prevention, personal communication with David Wiley, EPA, Dec. 8, 1999.

F7. Barrish, Robert. Delaware Department of Natural Resources and Environmental Control, Division of Air & Waste Management, personal communication, with David Wiley, EPA, Dec. 20, 1999.

F8. Zusy, Mark. Nevada Department of Environmental Protection, Bureau of Waste Management; personal communication with Craig Matthiessen, EPA, January 2000.

F9. Kleindorfer P., and E. Orts, "Informational Regulation of Environmental Risks. "Risk Analysis. Vol. 18, No. 2: p.155-170. 1998.

F10. Epstein, Lois. Engineer, Environmental Defense Fund, personal communication with David Wiley, EPA.

F11. Lawrence, Steve. "News of plot has suburban residents rethinking industrial neighbor." Associated Press, (December 6, 1999).

F12. Executive Summary, Risk Management Plan, Suburban Propane, Elk Grove, California. (December 27, 1999) www.epa.gov/enviro.

F13. Kovacs, D., G. Gibson, C. Chess, and W. Hallman. "Outreach Materials About Risk Management Plans: Guidance from Pilot Research." Cook College, Rutgers University Center for Environmental Communication. (September 1998).

F14. "Falling Through the Net: Defining the Digital Divide." Department of Commerce, National Telecommunications and Information Administration, July 8, 1999, revised November 1999, www.ntia.doc.gov/ntiahome/digitaldivide.

F15. "With Toxic Risk, Plans Vary," *Washington Post*, Oct. 10, 1999, p C1, C10-C11. "Plant Warnings Go Unheeded," *Washington Post*, Nov. 5, 1999, p A1. "Blue Plains Details Safety Plans," *Washington Post*, March 3, 2000, p B3.

F16. Lynn and Kartez, p. 517.

F17. Gablehouse, Timothy. Chair, Jefferson County Colorado LEPC, personal communication with David Wiley, EPA, Dec. 2, 1999.

F18. Rich, R., David Conn, and W. Owens, "'Indirect Regulations' of Environmental Hazards Through the Provision of Information to the Public: The Case of SARA, title III." Policy Studies Journal, Vol. 21, No. 1, 1993: p 16-34.

F19. U.S. GAO, "Environmental Information: Agencywide Policies and Procedures Are Needed for EPA's Information Dissemination," GAO/RCED-98-245. (Sept. 1998).

F20. Gablehouse, Timothy, Chair, Jefferson County Colorado LEPC, personal communication with David Wiley, EPA, Dec. 2, 1999.

F21. Lynn and Kartez, p. 515.

F22. Gablehouse, Timothy. Chair, Jefferson County Colorado LEPC, personal communication with David Wiley, EPA, Dec. 2, 1999.

F23. Howell, Kent. Georgia Department of Natural Resources, personal communication with Craig Matthiessen, EPA, November, 1999.

F24. Baram M., P. Dillon, and B. Ruffle, *Managing Chemical Risks: Corporate Response to SARA Title III*, Report to U.S. EPA, CEPPO, 1991.

F25. Lynn and Kartez, p. 515.

F26. Santos et al., "Industry Responds to SARA Title III: Pollution Prevention, Risk Reduction and Risk Communication," May, 1994.

F27. Davies, Terry and Jan Mazurek. "Industry Incentives for Environmental Improvement: Evaluation of U.S. Federal Initiatives," Resources for the Future, September 1996.

F28. *Guidelines for Technical Management of Chemical Process Safety*. Center for Chemical Process Safety, American Institute of Chemical Engineers, New York, 1989.

F29. Dick Knowles in Chemical Manufacturers Association Summary of the Kanawha Valley Safety Street Presentation, Charleston, West Virginia, June 1994.

F30. "Integrating Responsible Care, Community Advisory Panels, and Local Emergency Planning Committees into Successful RMP Communications to the Public," Nick Macchiarolo, Great

Lakes Chemical Corporation. Paper for CCPS conference 10/99.

F31. Ethyl Corporation, Health, Safety & Environment at www.ethyl.com.

F32. Baumann, J., P. Orum, and R. Puchalsky. *Accidents Waiting to Happen: Hazardous Chemicals in the U.S. Fifteen Years After Bhopal.* (December 1999): p. 12-13.

F33. D'Angelo, C. Factory Mutual Corp., personal communication with Craig Matthiessen, December 8, 1999.

F34. Stavrianidis, Paris. Factory Mutual Research Corp., "The 21st Century: Process Safety and Factory Mutual," Mary Kay O'Connor Process Safety Center Roundtable Meeting, June 2-3, 1999. http://mkopsc.tamu.edu/symposiums/briefing%20papers/21st_century.htm.

F35. The Social Investment Forum 1999 Report on Responsible Investing Trends, Social Investment Forum, November 1999, www.socialinvest.org.

F36. Gaskin, Russ. Managing Director, Social Investment Forum, personal communication with David Wiley, EPA, Dec. 3, 1999.

F37. Bateman, Mark. Investor Responsibility Research Center, personal communication with David Wiley, EPA.

F38. Howell, Kent. Georgia Department of Natural Resources, personal communication with Craig Matthiessen, EPA, November, 1999.

F39. www.aiche.org/ccps

F40. www.acs.org

www.ingramcontent.com/pod-product-compliance
Lightning Source LLC
Chambersburg PA
CBHW080640180526
45168CB00008B/3243